Waves of Fortune

Waves of Fortune

The Past, Present and Future of the
United Kingdom Offshore Oil and Gas Industries

DAVID UPTON

JOHN WILEY & SONS
Chichester · New York · Brisbane · Toronto · Singapore

Copyright © 1996 by David Upton

Published 1996 by John Wiley & Sons Ltd,
Baffins Lane, Chichester,
West Sussex PO19 1UD, England

National   01243 779777
International (+44) 1243 779777
e-mail (for orders and customer service enquiries): cs-books@wiley.co.uk
Visit our Home Page on http://www.wiley.co.uk
or http://www.wiley.com

All Rights Reserved. No part of this publication may be reproduced, stored in a retrieval system, or transmitted, in any form or by any means, electronic, mechanical, photocopying, recording, scanning or otherwise, except under the terms of the Copyright, Designs and Patents Act 1988 or under the terms of a licence issued by the Copyright Licensing Agency, 90 Tottenham Court Road, London, UK W1P 9HE, without the permission in writing of the publisher and the copyright owner.

Other Wiley Editorial Offices

John Wiley & Sons, Inc., 605 Third Avenue,
New York, NY 10158-0012, USA

Jacaranda Wiley Ltd, 33 Park Road, Milton,
Queensland 4064, Australia

John Wiley & Sons (Canada) Ltd, 22 Worcester Road,
Rexdale, Ontario M9W 1L1, Canada

John Wiley & Sons (Asia) Pte Ltd, 2 Clementi Loop #02–01,
Jin Xing Distripark, Singapore 0512

Library of Congress Cataloging-in-Publication Data

Upton, David
    Waves of fortune: the past, present and future of the United Kingdom offshore oil and gas industries / by David Upton.
      p. cm.
    Includes bibliographical references and index.
    ISBN 0-471-96341-0
    1. Offshore oil industry—Great Britain—Congresses. 2. Offshore gas industry—Great Britain—Congresses. 3. Continental shelf—Great Britain—Congresses. I. Title
    HD9571.5.U68 1996
    338.2'7282'094—dc20   96–11983
                                                                           CIP

British Library Cataloguing in Publication Data

A catalogue record for this book is available from the British Library

ISBN 0-471-96341-0

Typeset in 11/13pt Palatino from author's disk by Dorwyn Ltd, Rowlands Castle, Hants
Printed and bound in Great Britain by Biddles Ltd, Guildford, Surrey
This book is printed on acid-free paper responsibly manufactured from sustainable forestation, for which at least two trees are planted for each one used for paper production.

Contents

Introduction		1
1	The Early History of the North Sea	11
2	The North Sea and Politics	28
3	The Nerve-wracking 1970s, the Booming 1980s	54
4	Now – the Mid-1990s	74
5	The Scenario Planning Process	99
6	Technological Developments	117
7	The Way the Offshore Industry is Organised	141
8	Market Forces	163
9	External Influences	192
10	Four Scenarios for the Future of the North Sea	218
Appendix I	UK Governments from 1964 to 1995	233

Appendix II Major UK Legislation Relating to the Offshore
              Industry   235

Appendix III UK Oil Production, Expenditure and Export
              Statistics   240

Appendix IV UK Licensing Rounds, 1964–94   241

Glossary   243

Index   251

Introduction

The North Sea oil and gas industry came to Britain in the 1970s and 1980s as manna from heaven. It fed the nation's economy and sustained its hopes during one of the darkest periods of recent British history. The last traces of the British empire had finally gone. An energy crisis seemed set to last for ever. Wrenching financial upheavals finally reduced Britain from world leader to pauper, forced to borrow from the International Monetary Fund on the humiliating terms more normally reserved for third-world debtors.

To the British, the sea meant the English Channel – which had saved the country from invasion by Napoleon and Hitler – or the Atlantic, which came to symbolise an orientation towards the USA. The North Sea had always been the poor relation: a treacherous hunting ground for fishermen and the Navy, a bleak backdrop for the East Anglian or Yorkshire or Grampian coasts. But, from the 1960s onwards, waves of good fortune started to come out of the undistinguished North Sea and to overflow on to British beaches. As a result, the British are considerably richer now than they could ever have hoped to be 30 years ago.

At first, the new industry seemed far off and eccentric. I remember as a child watching workmen burning off the town gas in the pipes under the streets, so that the new North Sea gas could replace it. No one then had any idea of how much lay underneath British waters. I remember, as a Diplomatic Service officer newly returned from an overseas posting, the bleakness of a London winter during the "three day week" in 1974: we were only allowed electric lighting for three days, and by 3.30 pm on the other days the offices were dark and silent. The oil sheikhs, we thought, had us by the throats; I had returned to a country under siege.

Now the industry is accepted as fact of life: there is even a popular television series set on an offshore installation. We take it for granted that Britain is effectively self-sufficient in oil and gas, and that this will continue for some decades. We know that the platforms are there. When we fill up the car or build a factory, we may worry about the price of energy, but never that it might stop coming.

But industries grow, they become mature, and then they either develop or die. In the mid-1990s, the offshore oil and gas industry is experiencing its own revolution: it too is being forced to develop or die. The success or failure of these changes will profoundly affect the British economy, yet they are little understood outside the industry. Even within the industry, many specialists find difficulty in keeping up with developments outside their own fields.

The revolution is deceptively quiet. The same offshore installations are still out there, far away from the shores. Visitors to Great Yarmouth, or passengers flying from Edinburgh or Aberdeen to London by the route over Morecambe Bay, may catch an occasional glimpse of a steel structure above the sea; but for the most part the platforms are out of sight and out of mind. (How often does the television weather forecast show the thick arrows of heavy winds blowing out in the northern North Sea – and how few people stop to think what that means for the offshore workforce?)

For the most part, the same company names still appear proudly on the platforms, the logos adding a splash of colour to an otherwise dull structure. Rather fewer names now, especially in the share listings on the financial pages; but the large companies, the Shells and BPs, the British Gases and Essos, are still there and apparently the same. In some cases they still inhabit the grand buildings in central London or the City – though look inside the Shell complex or the BP tower and there are many empty or sublet floors. Esso is out beyond the M25 motorway, as are the exploration and production offices of British Gas.

The waves of fortune are still crashing down on to British beaches, feeding the terminals at Sullom Voe and St Fergus, Bacton and Theddlethorpe. But fortune will one day change. Britain continues to produce more oil than it needs, and almost as much gas, but it will not do so for ever. It is no simple matter to predict when the oil or gas will run out. They may run out for geological reasons; they may "run out" because we let them. Everything depends on the harnessing of a range of skills and resources, of knowledge and science and luck, without which the economic production of oil and gas is impossible. Everything depends on the economic and cultural expectations of society, and the ability of governments and oil companies to satisfy these, and to balance the conflicting claims of consumers, electorates and shareholders.

If this book has a hero, it is the industry itself and the people who work in it. No one studying the offshore oil and gas industry can fail to be amazed by the way that these people have faced up to the greatest technical and logistical challenges, by their ingenuity and ability, by their courage and imagination. Yet like every hero the industry has its darker side. In the same way that Sherlock Holmes was vain and dabbled in drugs, the oil industry is all too ready to believe its own propaganda and to succumb to senior management egotism. It trips on banana skins, some days it is unable to tie its own shoelaces, and it does not always help old ladies across the road.

If this book has a villain, it is all of us. Britain has consistently failed to understand the miracle of its natural wealth, or to take a balanced view of how to use it and how to control it. We have focused on small pieces of the jigsaw from time to time – most recently, the problem of how to dispose of offshore platforms that have outlived their usefulness. Tragedies such as Piper Alpha have reminded us of the risks involved. But for the most part we have turned on our gas fires and filled up our cars without considering where the energy comes from, or what we are doing to our children's inheritance, to our environment, or to our economic system. In our defence, this was often because no one could anticipate the size and extent of our oil wealth, or the technical solutions and problems that the years would bring. Governments are no more to blame than citizens. Politicians cannot create public debate where it does not exist, and at the same time they must respond to the issues (often "single issues") that do surface in the public mind. All too often the offshore industry has been out of sight and out of mind.

In facing the future, we need to give serious consideration to where the energy will come from, and to the balances that society must strike if it wants that energy to continue: between profit and idealism, between environmental anxiety and cheap fuel, between employment and efficiency, between flexibility and regulation. It is all too easy to sit back and wait for the "invisible hand" of Adam Smith to solve our problems for us.

The offshore oil and gas industry does not help its own case by being, in effect, a conglomeration of specialisations. Geologists, drillers, engineers, economists, financiers, environmental and safety specialists, and so on – each group speaks their own language and follows their own objectives. The latest industrial and technical developments are not difficult to find: they are well and promptly covered by a range of excellent industry journals, specialised magazines and newsletters. What is difficult is the task of interpreting each development, placing it in perspective, and gaining an overall view of what each development means, where each specialist is pointing. If this is hard for the insider, the person in the street (or on the back benches, or the trading floor, or in the lecture theatre) has even less chance. One objective of this book is to help the non-specialist reader to put the industry in perspective.

The second objective of this book is to think about the future of the industry and of Britain's largest single natural resource. Something this important needs careful planning and anticipation, at various levels. Companies involved in the industry must prepare their own forward plans, and it was at this level that the project began which led to this book.

The book arose out of a one-day workshop held privately at the Institute of Petroleum in London on 3 May 1995. The intention of the meeting was to assemble some of the most senior and respected figures from various aspects of the UK offshore oil and gas industries, and to spend a day looking 15 years into the future. To do this we used the techniques of scenario planning, and produced four "scenarios" – four possible futures – for the industry. The objective was to help all those involved in the industry by giving them a set of touchstones for their own planning processes.

The Institute of Petroleum is an independent body with over 8000 individual members and nearly 400 company members, with branches across the UK, and in Malta, the Republic of Ireland and the Netherlands. Its objective is to be the most respected independent European-based centre for the advancement of technical knowledge relating to the international oil and gas industry. It aims to enhance the professional standing of its members and to provide them with information resources to keep them up to date across the spectrum of industry developments and commercial affairs. We saw the Scenario

Planning Workshop as a useful way of carrying out this role. The scenarios were well received in the media, by Institute of Petroleum members, and by others.

However, the Institute was aware that the scenarios were a specialised publication and not immediately accessible to non-experts. We were also conscious that the "North Sea" offshore industry has reached a stage of radical change. Many people in the UK and elsewhere have a considerable interest in its future, whether they work in industries concerned with oil or gas, or have an interest in the British economy, or simply wish to form an idea of how long Britain's oil wealth (and the degree of independence it has sometimes given the country in international relations) will continue. Out of this came the idea for this book, which examines some of the issues and tries to lay out some of the evidence. (Although the original impetus was provided by the Scenario Planning Workshop, this book goes beyond it. The members of the Workshop are not in any way responsible for the author's judgements.)

Secondly, the Institute thought that others might be interested in a description of the "scenario planning" techniques we had used, and how these worked when applied to a practical question. They are an alternative to "fortune telling" – a way of grappling with a future which cannot be predicted with any degree of confidence. We hoped that other companies in the industry (or elsewhere) would use our scenarios or the method we had used in their own planning processes, even if only as a starting point.

A note on terminology will be useful. For 20 years, the British have spoken of "North Sea oil". This is inaccurate in two respects. First, Britain's hydrocarbon wealth includes gas as well as oil. When it comes to exploration and production (though not at the later stages of processing or refining and distribution, or in their economics) the gas and oil industries are very similar. Many offshore fields produce both oil and gas; most "oil" companies are involved in gas as well. Both are hydrocarbon products, formed the same way; their main uses are similar (e.g. heating, lighting, power generation and transport fuels) though they are by no means interchangeable. Gas has also contributed greatly to the British balance of payments over the last 20 years. Thus many references to the oil industry in this book ought to be read as "the oil and gas industry".

Secondly, oil and gas are now found in appreciable quantities in British waters other than the North Sea – in Morecambe Bay, for instance – and the relative contribution of the newer areas is likely to increase over time. The formal term for the offshore area in which the British government issues exploration and production licences for oil and gas is the United Kingdom Continental Shelf or UKCS. (The report issued by the original IP Scenario Planning Group was simply entitled *The UKCS in 2010*.) The subject of the present book might most concisely be defined as "the past, present and future of the offshore oil and gas exploration and production industries on the United Kingdom Continental Shelf". But old habits die hard, and to anyone who has lived through the last 20 years in Britain the phrase "North Sea Oil" has become a talisman of hope, a symbol of despair during the Piper Alpha tragedy, and the trigger to a well-rehearsed image of grey seas and lonely platforms. It is simpler than the full definition, easier to follow than the acronyms, and already a historic phrase. The author hopes that precisely minded readers will forgive him for using it from time to time.

The oil and gas industry abounds with technical jargon. Jargon terms have generally been explained where they are first used, and are also explained in a glossary at the back of the book. Where possible specialised terms have been avoided.

This book is divided into three main parts. First, and in order for the story of the "North Sea" to make sense to those (like the author) who have not lived and breathed oil for most of their adult lives, there is an outline of the history of Britain's oil and gas wealth and some idea of the techniques used to exploit it. Chapter 1 describes the gradual way in which these techniques were taken offshore and the early days of the British offshore industry – the period in the 1960s and early 1970s when it was not yet clear what could be found or, if found, exploited.

Once the size of Britain's windfall had been partly realised, five political problems arose, which have remained central ever since. Chapter 2 breaks from the chronological approach to examine each of them. The book then returns to tracing the history. Chapter 3 covers the 1970s and 1980s. At the beginning of these two decades, Britain oscillated between Labour and Conservative governments, with correspondingly different energy polices, while the finds grew larger and the investment began to pour in against a background of chaos in the world oil industry. During the 1980s, British oil production reached record heights under a Conservative government committed to letting market forces have sway. Then prices collapsed and the industry began to change: a change symbolised by the Piper Alpha tragedy in 1988, when 167 men lost their lives. After that there were cut-backs in production and falling oil prices. This period saw a dramatic reappraisal of the economics and technology used in offshore work, and the start of fundamental changes in the way the "North Sea" is run.

The fourth chapter tries to give an illustration of the state of the present-day offshore industry, although it is difficult to concentrate so much into a few thousand words. It tries to suggest perspectives rather than provide an exhaustive listing of activities (which can be found in standard industry reference books for those who need it).

Chapter 5 explains the scenario planning process and the way the Institute of Petroleum group used it. The chapter attempts to explain how scenario planning differs from fortune telling: it offers a tool for thinking ahead, rather than a "take it or leave it" prediction. It is a systematic process; one of the main steps is to identify "drivers" – forces or factors which are likely to have significant impact on change, one way or another, and then to try to think through the different ways that these drivers might develop. Using this method, we have tried to address the most critical uncertainties facing the industry, and to base our alternative views of the future on a careful consideration of the main issues.

Chapters 6–9 look in more detail at the "drivers" we identified. They are the areas, the changes, the activities that are most likely to shape the future. In this book, the "drivers" are divided somewhat arbitrarily into four chapters, purely for convenience in setting them out. Technological drivers (Chapter 6) looks at the speed with which exploration and production techniques will improve, the economics of using the new techniques and what they may make possible. Commercial and organisational drivers (Chapter 7) include the way in which industry organises itself – the size of the companies and the number of people they are likely to employ; the extent to which they can cut costs; and the changing relationships between different types of companies. Chapter 8 looks at drivers from the commercial environment, including the oil and gas price and the changing markets for both commodities; and the availability of capital and the management of risk; it also considers estimates of the size of the remaining reserves of oil and gas under British waters. Chapter 9 looks at external drivers, over which the oil companies have little influence: tax rates, the influence of the European Union, future patterns of energy use, the impact of disasters, and environmental pressures.

Chapters 6–9 do not pretend to predict the future; they try to give a survey of each "driver" area and to discuss ways in which it might develop. (These are the author's views, and not necessarily those of the members of the Scenario Planning Workshop.) Fifteen years is not long in many respects, and some of the latest developments may take half that time before they have their full effect. Other areas are totally unpredictable, such as the price of oil. This book tries to look at the implications of changes as well as at their possible direction and extent.

Chapter 10 reprints the four scenarios, as originally published by the Institute of Petroleum in 1995. Each scenario is set out as if by an observer writing in 2010. They have been slightly expanded to make them clearer to a wider audience. The purpose of these scenarios is not to "foretell" the future, but to suggest plausible alternative futures, both good and bad. The scenarios may never "come true". But they provide a background against which anyone concerned with the oil and gas industries in the UK can make their own forward plans. They give a concise overview of a range of possibilities and problems which a representative group of senior industry experts identified. They may make it easier to identify trends and spot "signals" of future developments.

The members of the Scenario Planning Workshop, with their positions at the time, were:

Mr Greg Bourne	General Manager, BP Exploration
Mr David Carr	Production Manager, Esso Exploration and Production UK Limited
Dr Rex Gaisford	Director, World-wide Development, Amerada Hess Ltd
Mr Alan Gaynor	Chief Executive, British Borneo Petroleum Syndicate plc
Mr Peter Kassler	Head of Group Planning, Shell International
Mr Ole-Svein Krakstad	Specialist, Improved Oil Recovery, Statoil
Mr John Mitchell	Head, Energy and Environment Programme, Royal Institute of International Affairs
Mr Terry Moore	Group Managing Director and CEO, Conoco Ltd

Mr Nick Perry	Vice President, Supply and Trading, Enron
Sir Hugh Rossi	Consultant, Simmons and Simmons (former MP and Chairman of the House of Commons Environment Committee, 1983–92)
Mr Colin Smith	Director, Oil and Gas, BZW
Mr Ian Ward	Director General, The Institute of Petroleum
Sir Ian Wood	Chairman and Managing Director, John Wood Group plc
Facilitator	Mr Peter Duff, BP Exploration
Chairman	Mr David Upton, Director, Stirling Reid Ltd

All those who took part did so in their private capacity. The results are intended to express a range of possibilities, and do not necessarily represent what anyone involved actually expects or wishes to happen in any particular area. The scenarios, and the companies referred to in them, are purely hypothetical illustrations of different ways in which the UKCS could develop over the next 15 years, depending on various factors and constraints.

The Institute of Petroleum, and I think all those who took part in the original Scenario Planning Workshop, believe that there is a real future for the offshore oil and gas industries in the UK. External factors such as regulation, the oil price and public attitudes will have considerable significance. Nevertheless, the industry, at all levels, decides its own willingness to innovate, to re-examine its own structures and working practices, and continually to seek more efficient, imaginative and socially responsible solutions to the challenges of the next 15 years. I hope that this book will contribute in some way to that process, by spreading ideas, stimulating debate and encouraging a realistic assessment of the present and future of the "North Sea" in those who will shape it or depend on it.

Any book such as this is a cooperative effort. It owes its inception, its plan and many ideas to those who took part in the scenario workshop. I am also grateful to the following for their help: Ian Ward, Peter Ellis Jones, Sarah Frost Mellor and Sjoerd Schuyleman of the Institute of Petroleum for their constant support and advice, and

Catherine Cosgrove and her colleagues in the Institute Library for providing answers to many difficult questions. My fellow director at Stirling Reid Ltd, John Stirling, for his industry contacts and expertise. Last but most of all, my wife Julia for her patience and support in this and so many other things.

I should add that the opinions expressed in this book are mine (or, in Chapter 10, those of the original Scenario Planning Workshop) and not those of the Institute of Petroleum or of anyone who has helped with the book. Any errors are similarly my own.

1
The Early History of the North Sea

FINDING OIL AND GAS

When the first drilling rigs moved out on to the waters around the UK, much of the technology they were using had already been developed and tested somewhere else. The oil industry was a brash, middle-aged operation with its own methods and its own language, and it reckoned it had the oil world fairly well sorted out.

Its infancy had been remote and primitive. Oil seeped from the ground in areas of central Asia and the USA, and the earliest oil wells were merely holes dug down a few feet to encourage the seepage. They were wide enough for a man to be lowered down in a basket. In the mid-nineteenth century the main value of "rock oil" (or its Greek alias, petroleum) was for dubious patent medicines, much the same as when Marco Polo saw petroleum in Asia.

The discovery that a liquid called kerosene could be refined from crude rock oil and used as an efficient and convenient fuel for lamps started off a series of enthusiasts and prospectors and eccentrics in America, land of the "Gold Rush". Where the natural seepages had been found, the prospectors started to drill – at first by hammering piles into the ground, and later with early forms of rotary drill. The first classic "strike" of all was in Pennsylvania; a retired railway conductor known as "Colonel" Edwin Drake found oil at a depth of 69 feet in August 1859 near Titusville. His find led to a small oil rush; it has been going on, to a greater or lesser extent, ever since.

Within 50 years, oil was to be a major industry in the USA and elsewhere; by 1900 it was used as a fuel for the new internal combustion engines, and for heating as well as for lighting. The oil industry in the USA was being rapidly centralised and dominated by John D Rockefeller and his Standard Oil company. Crude oil was discovered in new areas: in Russia, in the Dutch East Indies. By now the drillers could go down further: in 1902 the Spindletop well in Texas blew out from a depth of about 880 feet. By 1908 George Reynolds, drilling in what was then Persia for what became BP, could go down to depths of 1600 feet. In 1938, the first major Saudi oilfield was found at 4727 feet.

The early major finds were still made largely by guesswork. In 1902 the Texan Spindletop field was found the old way, thanks to the persistence of a local lumber merchant who had found natural gas bubbling up in local springs. He and his partner were regarded as insane by locals and professional geologists alike. (Many of the industry's heroes have been obsessed; we only remember those who were also right.) The Iranian oil fields were first sought as a result of oil seepages to the surface. Even in the 1930s American geologists exploring Saudi Arabia relied heavily on examining the surface geology, identifying structures which suggested that oil might be trapped below. Exact mapping, taking rock specimens and looking for fossils were the main ways of identifying and dating rock shapes. Aircraft were increasingly used for preliminary surveys of large areas.

But there was no guarantee that what was visible at the surface bore any relation to the earth forms below. (And, of course, surveys of the surface would pose special problems when the oil field was offshore and the "surface" was the bottom of the sea.) New methods were needed. Measurements of local changes in gravitational pull, or in the earth's magnetic field, were used from the 1920s onwards. They were, however, to be superseded by seismic methods, first used for oil exploration in 1921 in the USA, after the value of sound reflection techniques had been demonstrated during the First World War as a method of finding the range of enemy artillery.

The principle of a seismic survey is simple: a shock wave, or series of waves, made at the surface of the earth, is picked up on listening instruments some distance away. The shock waves travel down through the earth until they are reflected back to the surface. Different strata reflect in different ways. By analysing the varying times the waves take from source to receivers, a diagram of the slice of ground below can be built up. The technique is very complex and requires considerable precision in measuring elapsed times, as well as careful processing to turn a mass of data into an interpretable map. The skill and power of the processors are critical; after the Second World War the use of computers gradually made seismic methods into a more accurate tool. Using seismic, it is at last possible to glimpse below the surface; although the final and only reliable test is to drill.

Offshore seismic poses its own problems. Neither the source of shock waves nor the listening devices are usually laid on the sea bed. Instead, arrays of listening devices are trailed on the surface behind a ship, and the "shot" is a small explosion, also in the water. The accuracy of the results is of course much less, since the shock waves also have to travel through the water, which may distort them. A further problem, until navigational satellites became available, was

Plate 1 The only sure way to know what lies under the surface: geologists examining core samples from wells. (Photograph courtesy of British Petroleum)

that the seismic survey vessel could not accurately determine its position.

With these tools and others, the world's oil companies refined their ability to discover oil. As the oil fields were discovered, each one brought new problems: how to drill down to it, how to control the pressure of oil underground, and how to bring it safely to the surface.

The early development of drilling technology took place on land: in the USA, in Baku, Venezuela and the Middle East. Here, the technique known as fluid circulating rotary drilling was developed and refined. Diamond drill bits were first used in 1860; the rotary drill bit was designed by the father of the notorious Howard Hughes. As holes went deeper, hollow drill pipes made in convenient lengths were screwed together to keep on pushing the bit down. A fluid called "mud" was pumped down the inside of the pipe, helped to cool the bit and wash away the rock cuttings, and brought them back to the surface again. The familiar tower or "derrick" which stands over most oil wells is in effect a crane to raise and lower the growing length of pipe (with the bit, collectively known as the "drill string"). There is also a motor to turn the pipe and thereby the bit below, and a pump to move the mud. When the bit needs changing, the whole string has to be hauled out, each section unscrewed and put to one side, until the bit is brought to surface. Then the string is built up again and the new bit returned to the bottom of the hole. This laborious process is known as a "round trip" or "tripping", and the hard manual labour of unscrewing the string and manipulating lengths of pipe, as the derrick lifts them, is performed by the "roughnecks" or less skilled "roustabouts". A set of techniques, a range of devices and a whole vocabulary developed in the US oilfields are still in standard use all round the world. Many aspects of drilling can now be done differently, but the traditional methods have many advantages and are still widely used.

An oil or gas well is a complex construction activity. First, it requires a hole to be drilled through an unpredictable range of rock and other strata, and accurately aimed at the petroleum reservoir. The first benefit of fluid circulating rotary drilling is that the drilling "mud" keeps the drill bit cool during this process, and helps to circulate the rock cuttings back to the surface where they can be removed.

Secondly, drilling an oil well involves building a pressure vessel, similar to a gas cylinder or steam boiler – except that these pressure

Plate 2 Work on a drill floor (Photograph courtesy of Amoco (UK) Exploration Company)

vessels are built empty and carefully tested before they are used, while an oil well is built piece by piece as it is drilled, and is liable at any minute to receive the full pressure of oil or gas when it strikes a reservoir. The pressure found in reservoirs varies greatly, and even with today's techniques cannot be predicted with complete confidence. Early drilling attempts led to spectacular "blow-outs" as the top of the pressure vessel, or well-head, blew off, allowing the oil to shoot out of control to the surface, destroying the derrick and drenching the crew. In the early days, such gushers were a welcome sign to the prospectors. But offshore, where you cannot simply run away from the rain of highly inflammable crude oil, they are a nightmare.

The pressure vessel can also fail lower down: although it is drilled through solid earth, some strata are softer or weaker than others, and the sudden release up the well of oil held for millennia under high pressure by an impermeable layer of rock or salt may lead a softer layer above to fracture, causing an "underground blow-out". These are more complex, but often just as dangerous, since the oil and gas may still find their way to the surface near to the drilling rig.

The second benefit of fluid-circulating rotary drilling is that the fluid can be used to control these pressures down the well (or "downhole"). Once oil or gas at high pressure is encountered, it will naturally try to rise upwards through the spaces between the drill stem and the side of the hole, or through the hollow drill stem itself. But if the weight of fluid in these spaces is enough to balance the pressure, it will remain where it is, or can be brought gradually to the surface with the drilling fluid, or "mud", as the latter is pumped round. While the pressure is under control, cement can be forced down the sides of the well to make a permanent wall. As the well is cemented, proper well-head structures can be fixed in place to ensure that any likely pressure can be controlled. The "mud" that is used to control the well is a complex and expensive chemical. Some "muds" are oil based, others are based on water or synthetic fluids, but all types of mud can be "weighted up", by varying the proportions of heavy materials, to balance the pressure of hydrocarbons in the well. Pressure downhole is constantly monitored, and the driller responds to changes by adjusting the weight of the mud.

Balancing downhole pressure with mud weight is often referred to as "primary well control". If this fails, or if an inexpert driller loses control of the well, then the last chance of preventing a surface blowout is "secondary well control" – the so-called "blowout preventer" or BOP, a stack of valves or "rams" at the top of the well, which can close off the well quickly and contain it. For high pressure wells, BOP stacks may be rated to 15 000 psi or greater; very few wells are at such pressure.

Understanding events two miles down a nine-inch wide hole is a matter for experts. Pressures and temperatures may vary greatly, and are not easy to interpret. For instance, a sudden rise in pressure as a small pocket of gas or oil is encountered (called a "kick") will cause a rise in pressure in the mud above the pocket. This mud may fracture a weaker strata, allowing some of the mud to drain away – and the mud loss, or "lost circulation", will initially show up on the driller's instruments as an easing of pressure. If the well begins to behave unpredictably, there can be an anxious time for all concerned before it is brought back under control. A great deal of experience has been developed, first in the USA and later on elsewhere, in measuring events downhole and using this information to plan and control wells. In theory the behaviour of a well can be predicted, but even when there is enough information the mathematics are extremely complex. Drilling is as much an art as a science.

A further development, dating from the 1930s, is the ability to drill curved holes – so-called deviated or directional wells. At first, these were used from drill sites on land to reach out under the sea, reducing the need to go offshore. But they are invaluable for making the most use of a small artificial offshore drilling platform – which may have wells reaching out several miles to each side.

MOVING OFFSHORE

Drilling offshore is little different from drilling on land – except for two things. A stable and safe platform is required, together with regular supplies of mud, cement, drill pipe and power, not to mention water, food and accommodation for the drilling crew. All these are relatively easy and cheap to organise on land, and often extremely difficult and expensive to provide offshore. Secondly, if a well goes wrong on land – blowing out, or catching fire, or producing poisonous hydrogen sulphide gas – the crew can in the last resort run away from it. Offshore they are trapped on a small platform, surrounded by a sea which – in UK waters at least – is usually so cold that an unprotected person can only survive it for a few minutes. To take the drillers offshore was a triumph of engineering, economics, organisation and safety developments; though many of these triumphs were earned the hard way.

The industry's first steps offshore were tentative. Oil companies were tempted into the water by promising geological features, or simply by the unavailability of good prospects ashore. As early as 1910–20, wooden piled platforms were built in lakes in Louisiana and Venezuela, and drilling rigs mounted on them. Although some of these structures successfully produced oil, they were vulnerable to wood-burrowing organisms, and could only be built in sheltered waters. In the 1930s, the piled platforms moved tentatively out into coastal bays; but these were limited operations, in less than 20 feet of water, and usually less than a mile from the shore. For very shallow waters, drillers also used barges. These were towed out to the desired site and sunk to the bottom, where they acted as drilling platforms; later on they could be refloated and used elsewhere. Whether fixed or permanent, these platforms were serviced by drilling crews taken out by boats in the morning and returned at night to the shore.

Marine drillers did not lose sight of land until after the Second World War. The first piled rig to be constructed out of sight of land was built in 1947 by Brown and Root for the US independent Kerr McGee, nine miles offshore in the Gulf of Mexico. By 1950, it was becoming clear that major fields were likely to be found offshore in this area. Fleets of specially converted boats were used to supply the new platforms; in 1958 helicopter services began. Much of the impetus for development came from smaller companies operating in US waters, and unable to break in to the onshore fields largely controlled by the majors. But offshore drilling was also taking place elsewhere: in Saudi Arabia (the Safaniyah field, discovered in 1955) and Brazil.

Going offshore in the North Sea was an entirely different problem. True, the first steps in the Southern North Sea basin were in relatively shallow water. (Depths might be as little as 90 feet.) These were easily within the reach of the so-called jack-up rigs. These are effectively large square or triangular barges, with long legs at each corner. To move from well to well they ratchet their legs up into the air and float, like upside-down coffee tables. Once at the site of the well, they lower the legs down to the sea bed, and jack the hull of the barge out of the water to a safe height above the waves. Providing the legs have been properly sited on something solid, and the rig is strong enough, this provides a stable drilling platform. But the fickleness and ferocity of the North Sea came as a shock; a new generation of jack-up rigs, and later fixed platforms, had to be developed to withstand massive waves and strong winds. The "Ocean Prince", the first rig to discover oil (in non-commercial quantities) in the North Sea, was broken to pieces by a force eleven gale on the Dogger Bank in 1968, less than two years later.[1]

The stresses on these platforms, even in calmer water, were not fully understood. Rigs needed precise positioning to minimise the enormous strains on their legs. In 1965 BP's rig "Sea Gem" fell victim to metal fatigue, breaking a leg and falling over in relatively calm weather and 13 of the crew of 32 were killed; more would have died but for a passing cargo ship.[2] As well as better design and handling, improved means of assessing the "envelope" in which the platforms had to operate were required: better meteorological forecasts, and a better understanding of the nature and size of waves.

Another solution was the "semi-submersible". Again, the basis of this unit is typically a square barge, raised above two enormous pontoons. For transit, the rig floats on the pontoons. On reaching the well-site, buoyancy chambers in the pontoons are partially filled with water to sink the vessel until the pontoons are well below the surface, but still supporting the drilling deck high enough above the water. The pontoons below the surface minimise the effect of the waves. But "semi-subs" face their own problems: they have to be moored, and high winds or waves can break the moorings or tear them loose from the anchors. They are perpetually moving, but the sea bed they are drilling into is not. Steel drill pipe in long lengths in surprisingly flexible; but even when there are a few hundred metres of it between the drill rig and the sea bed, the drill rig can only move a few metres in any direction before the pipe will fracture, with possibly catastrophic consequences. If the rig breaks loose altogether and drifts it may collide with fixed installations in the more crowded offshore areas. Secondly, even the largest semi-submersibles are not immune to the effects of waves and swells. They have a curious, slow, rocking motion. In heavy swells they must compensate for the up-and-down movement of the rig, or else it is impossible to maintain a controlled pressure in the drill bit.

When exploration moved into the northern waters of the North Sea, and water depths of several hundred feet became common, the jack-ups were no longer able to operate and the semi-sub came into its own. Once oil or gas had been found, it was common to erect a permanent structure, or set of structures, over the field. The construction of these artificial islands, so vast on land and so tiny in the sea, demanded new techniques and even new dry docks. And of course, once oil is found offshore, it must be transported to somewhere where it can be used. Tanker loading facilities, offshore storage tanks and later pipelines had to be developed.

Last, supplying the offshore platforms and rigs, and moving their crews out and back, required vessels and helicopters that can function and navigate reliably in bad weather. If supply is cut off, the platform must be able to support itself. If there is an emergency, it must provide its own fire-fighting and rescue capability, since shore-based emergency services cannot reach it in time to make a difference. The small stand-by vessels, required by law to attend constantly on each platform, are unable to provide more than limited assistance; their role is primarily to pick up any crew member who falls in the sea. In the worst weather, these vessels are arguably at more risk than the platforms they guard.

The oilmen came to the North Sea with a lot of their exploration and drilling technologies ready made. British waters forced them to rethink and redevelop many of their marine skills. They were also to stretch the commercial ingenuity and the diplomacy of the oil companies.

THE EARLY DAYS OF THE NORTH SEA: EXPLORATION AND GAS

It would be convenient if the history of the North Sea fell into neat periods. The nature of the industry, of the British economy, and indeed of popular sentiment, precludes this. Between finding an oil or gas field and exploiting it there is always a lag – usually years, sometimes decades. Changes in the economy, too, often take long periods to work through. Changes in popular beliefs, the "accepted wisdom" which underlies democratic politics – and the conformist majority of investors and industrialists – are no quicker. But the industry, and the country in which it operates, have changed dramatically since the first gas find in 1965. A central argument of this book is that, in the mid-1990s, the industry is undergoing further radical changes – and that these are still not apparent to the majority of people in the UK whose lives are affected.

Oil was first discovered in commercial quantities in the UK in 1938, at Eakring near to Sherwood Forest. The discovery was kept secret. During the Second World War, onshore exploration and production was expedited, in great secrecy; at its peak in 1943 the UK produced 3000 barrels per day from 106 wells.[3] The 1934 Petroleum Production Act vested all petroleum reserves in the Crown, rather than leaving them as the property of the "surface owner", and set out conditions for licensing their exploitation.

Overwhelmingly, however, Britain's energy in 1945 came from coal. Homes were heated by it, factories ran on it. Even its main rivals, electricity and manufactured "town gas", were mostly derived from inefficiently transforming the energy of coal into another form. Oil was used largely for transportation, and it was imported. In 1950, coal supplied 90% of primary energy requirements. But by 1966, this pattern had changed: imported oil now accounted for 40% of primary energy use.[4]

During the post-war years, as the country grew more dependent on oil, Britain faced two serious threats to its ability to import what was becoming its life-blood. The first was the nationalisation, by the Mossadeq government, of BP's interests in Iran in 1951 – interests which had been the guarantee of oil supplies for the Royal Navy since before the First World War. The nationalisation was reversed in 1953, with help from the CIA and MI6, but the Iranians won new terms. BP (then called Anglo-Iranian) was replaced by a consortium of companies; it was felt that the British monopoly was no longer defensible as Iranian nationalist opinion grew. BP had 40% of a consortium that now included Shell, five American majors and even a French company, CFP. The terms of the concession were rewritten, and the old agreement under which a British company effectively owned and controlled the Iranian fields had gone for ever. The second problem was the Suez crisis in 1956, which temporarily cut the main shipping route along which imported oil reached Britain.

The real threat that was to emerge was still disguised. Oil prices remained low. Despite events in Iran, this was still the era of the seemingly impregnable agreements between the "majors" and the producing countries, of which that between Aramco and Saudi Arabia was the largest. British fears were directed at our developed world competitors. The lack of US support over Suez, and resentment that the Americans had been allowed to take a share of Iranian oil, was one of many tensions underlying the "special relationship" which Britain claimed to have with the USA. American analysts also seemed to ignore the real issues, and to anticipate that their profitable concessions would go on for ever and that the terms of oil trade would stay firmly in favour of the consumers, and in particular of the US major oil companies who acted as middlemen.

The first confirmation that there might be large quantities of oil or gas under the North Sea came from the Netherlands, where in the late 1950s the Nederlandse Aardoljie Maatschappij (NAM – a joint venture by Shell and Esso) discovered the enormous onshore Groningen gas field. The geology of this field was similar to that of the southern North Sea basin; it was a fair guess that similar fields might exist in British waters too. Very small gas finds had indeed been made before the war in Britain.

The first offshore commercial gas finds in British waters were made by BP in November 1965 – what was subsequently to be the West Sole gas field. Sadly, the exploration rig "Sea Gem", which had made the find, sank six days afterwards, with the loss of 13 lives. Other finds followed, their names a strange mixture: Leman, Indefatigable and

22 *Waves of Fortune*

Plate 3 The first North Sea platform: BP's platform "A", West Sole gas field, in 1967 (Photograph courtesy of British Petroleum)

Hewett. They were quickly exploited; by 1967 natural gas was being used in British homes. The Leman field became the world's largest offshore gas field in production. A revolution was about to begin.

The gas industry was a closed world that had relied for many years on manufactured "town gas". Town gas was not a primary energy source, and not seen as strategically vital. It required special handling and installation, and it smelt. (Pure natural gas does not have a strong smell; it is artificially "stenched" with mercaptans before delivery so that leaks can be detected.) In 1967, gas accounted for only a small part of Britain's energy requirements; but within 10 years, North Sea gas was to supply a quarter of primary energy.

The revolution was slowed down by the limited market, and by disagreements over contract terms. Gas is not traded internationally in the same way as oil. It is difficult to transport and store, so it is usually consumed near to where it is produced. As a result of this, and of legislation, all gas from UK waters had to be sold into the UK market via a single purchaser, the British Gas Council (later the British Gas Corporation). The contract negotiations in the late 1960s and early 1970s between the Gas Council and the operators were complex and left a legacy of bad feeling.

From the Gas Council's point of view, this was a new fuel. It had to spend heavily to develop a market for "North Sea Gas" in the UK, particularly to build pipelines to transport it and to convert networks that had handled the older "town gas". It had to challenge the traditional domination of coal; it expected to achieve only a gradual success. It negotiated contracts with these problems in mind.

Gas contracts are more complex than those for oil. First, the demand for both oil and gas fluctuates, but crude oil undergoes several processes between well and end-user. Fluctuations in final demand for gasoline and kerosene are ironed out while the crude is being transported in tankers and processed "offline" in refineries. By contrast, most gas undergoes only very limited processing or storage between well and final customer: in effect it is produced "online". Typically, gas contracts have been written on the assumption that gas will be produced only as needed. The offshore gas platforms will act as taps, turning the supply up or down as the customer requests. Secondly, because the market for UK gas was limited to a single country, and there was more gas available than could be absorbed, the sole purchaser was able to control the long-term rate of production.

Prices were only one part of the contracts. The initial prices negotiated by the Gas Council in the 1970s were relatively low (2.89 cents per therm for Indefatigable gas in 1968, or 3.5 for Viking gas in 1971, compared with Dutch prices of 3.7 in 1969 and 3.85 in 1970[5]). Even though the contracts reportedly included price escalation clauses, the limited increase in the retail prices of gas in the UK during the 1960s and 1970s was not comparable with the windfall profits earned by oil fields after the 1973 oil price rises. But that is with hindsight. Other conditions of the contracts were just as important, and as disappointing for the companies.

First, the Gas Council, most of whose sales varied with seasonal and weather factors, demanded a considerable range between the average, maximum and minimum daily contract quantities. The ratio of the average daily quantity to the maximum that may be required on any one day is known as the "load factor". A low load factor (say 60%) means that the operator has to accept a wide fluctuation in production: it must build facilities to handle the maximum contractual throughput, but accept that they are likely to handle a far lower volume most of the time. (A high "load factor", say 80–90%, implies a much smaller fluctuation in production rates.) The Gas Council's low load factors put up the development and operating costs for the companies, but reduced their revenue.

Secondly, the Gas Council insisted on slower exploitation of the fields than the companies planned. Several large gas fields were coming on stream in the 1960s and the Council did not believe that it could market their entire unrestricted production. For the companies, slow development is a disadvantage: by delaying the receipt of revenue, it lowers the ultimate value of a project and the rate of return to the company, in some cases below what is acceptable to them.

The negotiations were complex and heated, as much between the operators themselves as with the Gas Council. In the end, the operators had little choice. They were not allowed to sell their gas elsewhere; in theory sales outside the UK could be authorised by the government, but an attempt to sell Viking gas to continental buyers was disallowed. The major gas fields – Indefatigable, Viking, Leman, Hewett – had been discovered and were producing by 1969. Exploration in the southern basin fell off relatively quickly after then, presumably because companies were disappointed with the contracts they had been awarded, and there was so much gas potentially available that the market seemed unlikely to pick up. Some fields (such as Anglia) which were discovered in the early days were not developed until 20 years later when the market had changed entirely.

Surveying continued in the central and northern North Sea, this time for oil. It was touch and go. By 1970, 25 "wildcat" exploration wells had been drilled; and 21 of these were dry.[6] Several rigs had been damaged, some critically. However, in that year the tide turned. Sir Eric Drake, the Chairman of BP, who made the unfortunate statement (in April) that there would not be a major oil field in the North Sea, was announcing only six months later the discovery of the Forties field, which was calculated even then to hold a billion barrels of oil. (An underestimate, but understandably so: at the time Britain's total oil production, from onshore fields, was about 2500 barrels a day.) Forties was one of the two largest fields so far found in UK waters, and a giant by any standards. The race was on.

A first priority for the British government was to determine which country owned the waters. Historically, states had been more concerned about 12-mile limits around their coastlines. For strategic reasons, they needed the right to control waters from which long-range naval guns might bombard their shores. No one ascribed any economic value to the "high seas", except as a highway; they were "no-man's-land", and British policy had traditionally encouraged freedom of navigation. The UN Law of the Sea Conference, in 1958, had adopted a Convention on the Continental Shelf which provided a framework by which states could claim much greater rights over nearby waters. Britain took advantage of this; the Continental Shelf Act of 1964 gave effect to the UN Convention and made provision for the exploration and exploitation of offshore petroleum. (The timely annexation by Royal Marines of the uninhabitable island of Rockall, way out into the Atlantic, considerably extended the area claimed by the UK.) Over the next 10 years, bilateral agreements between the countries surrounding the North Sea finalised the international boundary lines.

Britain was in a strange position in the 1960s. It had fought a world war and impoverished itself in the process. It was rapidly losing the empire on which so much of its wealth and self-respect depended. Despite Marshall Aid and national sacrifice, the late 1940s and early 1950s had been even more austere than the Second World War itself. Bread, for instance, was rationed only after the war ended. Conscription to the armed services continued to affect most young men.

With the "swinging sixties" came a new mood of national optimism: worldwide cultural phenomena such as the Beatles, and in 1964 the return of a Labour government under Harold Wilson. But the old problems remained: the set of national beliefs and sense of purpose that had sustained Britons, from Kipling to the Battle of Britain pilots, was withering. Britain, a perceptive American said, had lost an empire but not yet found a role. An attempt to join the Common Market was vetoed in 1963 by General de Gaulle. Attempts to maintain the "special relationship" with the USA produced some results, but were never entirely comfortable for either partner. The death agonies of empire included the Rhodesian Unilateral Declaration of Independence in 1965. This overshadowed much of Harold Wilson's first period in office. It also helped to muddy relations between the British government and BP and Shell, when the oil companies were found to be supplying Rhodesia indirectly, in defiance of the United Nations economic sanctions imposed at Mr Wilson's request.

British industry was badly managed, and hampered by poor industrial relations; the country was firmly set on a course of declining productivity. Industrial structure and technology failed to develop, and in one industry after another Britain lost what lead it had possessed before the Second (and sometimes only before the First) World War. One of the energy industries in particular came to symbolise bad management and poor industrial relations: the history of the late 1960s and early 1970s saw repeated confrontations with the coal miners, leading to the 1973/4 crisis when the Heath government put industry on a three-day week in an attempt to conserve coal in the face of a miners' strike, and the National Union of Mineworkers eventually brought down a government.

In 1967, after the devaluation of sterling, the International Monetary Fund (IMF) had forced the Labour government to impose unwelcome monetary targets. The economic malaise which affected the country in the 1960s and 1970s was so great that an outside investor – say an American oil company – could have been forgiven for drawing parallels with many of the third-world countries where the company had traditionally invested: weakened governments, a shrinking and largely old-fashioned industrial base, dependence on the USA and vulnerability to external economic conditions. National pride clung to symbols (such as the Royal Family) and to the past.

The political climate was scarcely reassuring. There was a large gap between the traditional owners of the country's wealth (a Tory party still dominated by wealthy aristocrats like Macmillan or Douglas-Home) and the left (a Labour party still committed to nationalise the "commanding heights" of the economy, and whose previous spell of government had led to the most far-reaching, and expensive, social reforms ever seen in the country). Yet it was from these unpromising and tentative beginnings that a major industry grew – one which was to transform the British economy beyond all expectation.

REFERENCES

1. B Cooper and T F Gaskell, *The Adventure of North Sea Oil*, Heinemann, 1976, p 120.
2. Cooper and Gaskell, *The Adventure of North Sea Oil*, p 131.

3. C Reader, Secret goings-on in wartime, *Petroleum Review*, September 1991.
4. G Arnold, *Britain since 1945*, Blandford, 1989, p 96.
5. J D Davis, *High Cost Oil and Gas Resources*, Croom Helm, 1981, p 132.
6. Cooper and Gaskell, *The Adventure of North Sea Oil*, p 72.

2
The North Sea and Politics

As it became clear that there were significant quantities of oil and gas under the North Sea, five areas of political sensitivity soon emerged. They have remained difficult areas ever since, and this chapter steps back from the chronological approach in order to examine each in turn.

First, how are the profits from exploiting the resources to be divided between the government and the companies? (With a subsidiary question: which government is entitled to the state share – Westminster? Brussels? or a Scottish national assembly?) Secondly, how should the oil wealth be used – and, as part of this question, should it be produced as fast as practicable, or should production be rationed in the interests of some broader economic plan?

The third issue has revolved around the offshore workers: their safety, their conditions, and their ability to organise in trade unions. The fourth question has been the degree to which national industries have benefited from the increased activity – as operating companies, as providers of services and supplies, as fabricators – and the numbers of jobs created for Britons as opposed to foreigners. The fifth area has been the effect of the industry on the environment.

In one way or another, these five problems have shaped the development of the industry.

SHARING THE WEALTH

The oil or gas finds represented an enormous source of income. The state has the power to license or deny production, and to set whatever terms it chooses. It wants to claim a share of the wealth for its people: this is sometimes referred to by the British government as capturing the "economic rent" of the resources. However, production cannot take place without expertise and capital. If commercial participation is sought, then the state must offer terms which are attractive enough to bring in the companies and to enable them to raise money. All oil-producing countries have faced the problem of striking the right balance between protecting national interests and attracting external funds and participation.

In the 1930s, the Middle Eastern producers had sold what turned out to be very generous concessions to the major companies, who were allowed to produce oil cheaply at low local tax rates and with considerable control over their own operations. Realising their mistake, the producing countries adopted various tactics to change this position in the period between 1945 and 1973. They gradually raised the national share of the profits, and developed their own expertise in order to understand the companies' operations and profit/cost structures, in order to strengthen their own negotiating abilities. They formed OPEC, the Organization of Petroleum Exporting Countries, in 1960, in an attempt to form a united front. In some cases, they tried to nationalise the oil and gas fields outright – sometimes successfully, sometimes not. The oil companies clung on to favourable agreements as best they could, but the pressures had continued to mount and the dramatic oil price rises of 1973 and later years were, with hindsight, inevitable. The quadrupling of the oil price in 1973/4, and subsequent rises, had a radical effect on the world's economy. In Britain, they led to inflation, involuntary transfers of wealth both within the country and from it to outsiders, and much uncertainty.

Successive British governments came to the problem of how to control UK oil production with a strong set of preconceptions. Britain is a western country, with loyalties to its partners. It is a developed economy, which possibly suffers more than it gains from high energy prices. World economic growth is in Britain's interest, as the British economy is unusually sensitive to external factors; world growth also suffers from high oil prices. Battered by Nasser over Suez and by Mossadeq's attempt to nationalise the BP oil fields in Iran, Britain did not feel any natural alliance with the developing world oil producers, nor sympathy with their methods or their regional political agendas. Boycotts, outright nationalisation and other politically maverick behaviour were not for the majority of parliamentarians at Westminster. For one thing, Britain has extensive foreign investments; it protects these through good relationships and by encouraging the rule of law.

The North Sea itself posed particular economic problems. Even when operated in the most efficient manner, it is a high-cost producing area compared with giant onshore fields in the Middle East, where the costs of production may be less than $1 per barrel. Enormous investment in North Sea infrastructure was required: as much as a quarter of total UK industrial investment went into the North Sea in some years. This money had to come from somewhere, and the major oil companies were well placed to provide it – especially in the early years, when the British government was reduced to asking the International Monetary Fund (IMF) for special credits to prevent sterling from collapsing. The oil companies also had the skills necessary to venture offshore. While Britain could in theory have built up a national oil company to handle all exploration and production, it would have taken time to acquire the technical and commercial experience, even if the finance could have been found. BP was the national champion, but could not handle the whole North Sea by itself. In one way or another, Britain needed the international oil companies.

But increasingly the companies needed the North Sea, and Britain found it had some negotiating advantages. It was soon clear that some very large oil fields, such as Forties, might be available, and these are the ones where the best profits are traditionally made. Large fields are needed to sustain supplies to large marketing networks. The existing large fields in the Middle East and elsewhere were already under development, and already licensed. Ambitious crude-short companies saw the North Sea as a possible salvation. Secondly, after 1973, oil and gas supplies under UK control became a strategic asset: an alternative to OPEC for any governments or companies anxious to have reliable sources of supply. The result was a complex bargaining process between companies and the British government, in which neither side had the upper hand.

Even when it has been decided how the wealth should be shared, it is technically difficult to achieve a precise rate of government take. Oil (and gas) fields vary in their economics. Early indications of size may be very deceptive, yet heavy investment is often needed before the final profitability of a field can be known. Typically, most of the capital investment is required in the first two to five years of an oil field's life; but it may take two or more years before the first oil is produced commercially and any return is earned. Once the wells have been drilled, the production platforms built and the pipelines and other infrastructure set up, however, the operating expenditure is relatively low, and revenue begins to flow in as the oil or gas is lifted and sold.

Governments want to ensure that companies do not make excessive profits out of fields during the most profitable stages. The companies may provide capital, labour and technology, and be entitled to a return on these three factors of production. But the nation is providing the natural asset, and is entitled to a return or "economic rent" for that.

In practice, such economic theory gives way to horse-trading. Taxes have in effect been negotiated with the industry, before being imposed unilaterally. The main question has been how much the state can take without driving away investment. Much of the horse-trading is about perceptions of the horse's future ability to win races: the companies (separately, or represented by a committee of the UK Offshore Operators' Association) pointing to the costs and problems of their fields, the governments to the rewards.

The argument is about a string of expenditures and revenues that may spread out as far as 20 years ahead. In order to express these future cashflows in a way that can be easily comprehended and compared, investment analysts assess the "time value" of money. Money now is worth more than the same amount of money in a year's time (because money now could be earning interest for that year). So the value of a future payment can be "discounted", against an assumed interest rate, to give a net present value or NPV of the payment. Discounting all the components of the future cashflow of a field – 20 years' estimated expenditures and receipts – gives a discounted cash flow (DCF) or internal rate of return (IRR) for the field: this is a cashflow statement that takes account of the time element. Negotiations about taxes have typically revolved around sets of calculated IRRs for "typical" fields. For a range of reasons, it is impossible to predict an IRR with certainty (for one thing, it is impossible to say what the interest rate will be for the next 20 years). Assumptions about the IRR, however, underlie many negotiations between government and companies.

Inevitably, the solutions reached have been complicated. It is usually recognised that companies should be allowed to offset some part of their initial development costs against some or all of their taxes. Tax concessions made to compensate for initial capital expenditure engage the arts and skills of accountants on both sides: what is allowable, how the expenditure is defined, against what other income the expenditure is allowed, and so on. The first attempts to set rules were too generous. The 1973 report of the Public Accounts Committee pointed out that the companies were paying very little tax indeed; they had been allowed to offset against North Sea profits not only capital expenditure in the North Sea, but also tax losses accumulated by other activities.[1] After this, tax rules were tightened and complex "ring fencing" provisions introduced.

Tax rules quickly begin to have an effect on the activity that is being taxed. If exploration and development costs are partly paid by the government (in the form of a tax allowance against other income) companies with other taxable income may be more willing to develop new fields than otherwise; the tax considerations will factor directly into their IRR calculations. A company with no taxable income against which to offset expenditure may take a less enthusiastic view of the same project. The former may invest in facilities to ensure the maximum recovery from a field; the latter may go for a quick, cheap, minimal system to maximise its cashflow even if the field reservoir is inefficiently depleted. The former may drill several "wildcat" exploration wells for the same net cost to its balance sheet as the latter's drilling one. Any tax rules risk skewing development away from what is economically or technically sound to what is tax efficient.

Companies are taxed in various ways. First, all companies operating in Britain pay Corporation Tax on their profits, and make National Insurance contributions. For most British companies, these are the only direct taxes they pay. Secondly, oil companies have also been obliged to pay additional levies, say on each unit of oil, or royalties. If the oil price rises dramatically, oil in their hands will multiply in value, and they may be expected to pay a "windfall tax". They may be required to hand over a percentage of the oil or gas they produce to the government. In Britain, these options have led to a range of special taxes: SPD (Special Petroleum Duty: 1981–2); PRT (Petroleum Revenue Tax: 1979 to the present, but now being phased out) and APRT (Advance Petroleum Revenue Tax). These taxes are complex and the accountancy profession has found great difficulty in

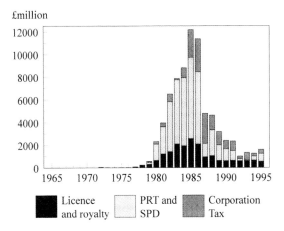

Figure 1 UK Tax Revenue, by Tax Type (Source: DTI *Energy Report*, 1995)

applying them consistently in company accounts. (KPMG Peat Marwick said in 1990 that "the failure of the accountancy profession and the industry to produce a consistent and comparable procedure for PRT accounting is disappointing".[2]) Thirdly, oil companies have to pay for the field licences themselves: initial fees for granting the licence, and periodic rental fees.

The graph in Figure 1 shows the different amounts of revenue collected by the three types of tax. Typically, licence fees have contributed relatively little. In times of high oil prices, taxes on petroleum produced (which are now being phased out entirely) have contributed the most. Taxing the companies only through the normal Corporation Tax would not make sense in times of high windfall profits.

Gas taxation has been different. In earlier days, the government may have felt that the state was capturing the "economic rent" through the contract terms with the state-controlled Gas Council. In 1981, as British Gas was being privatised and its role changing, a "gas levy" was introduced.

Once a pattern of taxation has been set it develops its own momentum and becomes very difficult to alter. Consistent treatment by the tax authorities is important to the oil and gas companies. They can evaluate a prospect given existing tax rates, and make a decision whether to go ahead. What planners find difficult to cope with are sudden changes in tax rates, which can play havoc with their carefully calculated IRR. A good track record of consistent tax treatment is an advantage for any country wishing to attract international investment to its oil fields. However, as the UK government found when it ended PRT, well-judged changes may attract considerable criticism, but ultimately be accepted as beneficial in the long run (see Chapter 9).

In order to have an effective tax policy, and to retain a proper share of wealth for the nation, the government must have a degree of control over the nature and extent of the companies' activities. There are two ways that governments normally control activity in oil or gas fields. They can operate various forms of production-sharing contracts, in which the government provides the acreage, the company offers the capital and the expertise, and the two share the oil or gas produced – if any – according to an agreed formula. Or governments can license the field to a company, retaining long-term ownership but allowing the company the exclusive right, subject to agreed conditions, to exploit the field for a specified period. This is the pattern chosen in the UK.

British licence terms vary in the degree of control and security of tenure they offer. They are issued for specific "blocks" or areas. Licences have been issued for as many as 46 years and as few as 18 (with a qualified possibility of a further 18). Initial payments have varied from £25 to £390 per square kilometre (the average block size is roughly 250 square kilometres) with fees for subsequent periods escalating up to maxima between £29 and £6750 per square kilometre. Work obligations are imposed on companies. To obtain and keep the licence, they must agree and carry out specific programmes: shoot so much seismic, drill so many wells, etc., within a prescribed time. (This is to prevent companies simply sitting on licence areas to keep their competitors out, although some would argue that it has not been entirely successful.) The work obligations of each licence are individually negotiated and vary greatly. Licences contain limited relinquishment provisions, forcing companies to hand back title to part of the licensed area after a certain time. Once again, the effect of these has been controversial.

Licensing "rounds" have been held on average every two years. (Although a small number of direct applications for blocks are considered, in practice companies must wait for a licensing round and can then bid only for those blocks that are on offer.) The choice of blocks on offer is up to the government, which allows it to focus attention on specific areas. (For instance, the seventeenth round, for which applications must be submitted by November 1996 and for which awards will be made in 1997, offers 68 "tranches" (groups of blocks) of which 41 are in the Rockall Trough, a "frontier" area in the Atlantic.) The numbers of blocks on offer has fluctuated over the years: the second round (1965) offered 1102 of which 127 were taken up; the 6th (1978/9) offered only 46, of which 32 were taken up. (See Appendix IV for a list of the rounds.)

All these variables, together with the uncertainties of the production process and the economics of the oil industry, were new to the British government. It took some time to build up expertise. There was little popular political debate on the subject, largely because few understood it. At least Britain avoided the mistake made by Denmark: all Danish waters were licensed to one company, the A P Moller company, in 1962, with almost no conditions attached. The Danish government spent the succeeding years trying to claw back what it had given away. The British licensing system in 1995 is substantially the same as it was in the first licensing round in 1964, although it was altered for a while in the late 1970s. Behind the legal framework, however, a second pattern was set: consultation between government and companies, even during the most antagonistic periods. Given the complexity of legal and taxation issues, it seems to make sense to avoid dogmatic adherence by either side to formulae or principle, and to leave room for compromise and creative tension.

CONTROLLING DEPLETION OF THE FIELDS

In the 1970s, there was no time for a serious debate about how Britain's national oil wealth should be used. There was some discussion about whether the money realised should be used to fund tax cuts and promote private enterprise, or whether it should be used more directly to fund state support for British industry, or even whether it should be handed out in equal shares to every British adult as a "North Sea stock".[3] But this was overshadowed by the urgent demand for every barrel of oil that could be produced.

It is possible for governments to control the rate at which oil and gas resources are exploited. They may be extracted as soon as possible, or they may be held back and spent gradually as part of some larger economic plan. This was done, for the gas fields, by the Gas Council; as we saw in the previous chapter, it used its monopsonistic (or sole buyer) position to slow down the development of the big gas fields. (It is interesting that the liberalisation of UK gas markets has now led to a gas oversupply, despite the greatly increased market size. Because the market is still only a national one, oversupply has caused price collapse and postponement of projects.) For British oil, however, there are no such demand limitations. British production is not a large enough share of world production to affect the world supply/demand balance; effectively every barrel produced in UK waters can be sold without affecting the price. This would not apply, for instance, to Saudi Arabian oil.

Careful decision on the timing of oil exploitation was a luxury that Britain was never able to afford. In the mid-1970s, when production was starting and a debate on extraction rates might have taken place, the world was in the grip of the "energy crisis"; the only pressure, and the only politically viable course, was to produce as quickly as possible. Emergency legislation, the 1976 Energy Act, had to be introduced to limit the use of energy: all available sources were desperately needed to prevent an unacceptable change of lifestyle in the country, and to limit the power of the coal miners' unions.

The evidence suggests that any early debate about exploitation of the fields would have been based on the wrong assumptions. Early "prophets of doom" were mistaken. Aubrey Jones warned in 1981 that "The United Kingdom could become a net importer of oil before the end of the 1980s unless substantial new finds are made and developed. Thus the United Kingdom could become dependent on imports by the late 1980s at a time when oil supplies from OPEC and elsewhere may be neither secure nor sufficient. Such a dependence could be avoided only through a depletion policy extending through the 1980s."[4]

Jones's fears were wrong: substantial new finds were made; known fields were found to be larger than expected. Reserves below the UKCS waters are estimated each year by the Department of Energy (now part of the Department of Trade and Industry, or DTI) and published in what is usually known as the "Brown Book". Estimates are made of three categories of identified reserves: proven, probable and possible. A fourth estimate is made of "undiscovered" reserves; this is based solely on the statistical likelihood of further reserves being found, given what is known about surrounding areas. It is obviously the least reliable and is usually expressed as a wide range of values. All four estimates have continually increased; in 1995

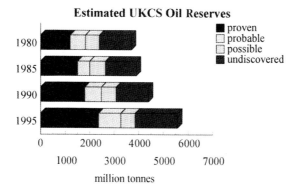

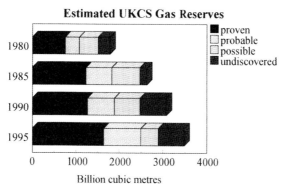

Figure 2 Estimates of UK Reserves (Source: UK Department of Energy/Trade and Industry estimates)

Britain is still some way from being a net importer of hydrocarbons (see Figure 2).

All UK governments have allowed production and development at high rates, because of perceived energy shortages, because they needed the money, or because of an ideological commitment to free market forces. A good statement of this commitment is made in the 1982 report by the House of Commons Select Committee on Energy, on "North Sea Oil Depletion Policy". The committee was dominated by Conservative members, and its report argued that it would be futile and self-defeating if the government tried to "substitute its own judgement for that of the oil companies in an attempt to override the technical, political and economic uncertainties with the aim of bringing about a desired production profile in the North Sea".[5] However, the Committee also saw a role for government in selectively encouraging the development of "frontier" areas through licensing rounds, maintaining a stable and encouraging environment for investment, but retaining reserve powers to control production in case of national emergency.

Given future strains on the public purse, such as the anticipated overhang of pension and social security funding requirements in future years, it is open to question whether any British government would be able deliberately to slow down the rate of oil production. However, as the remaining reserves in the UKCS eventually come nearer to an end, it is possible that the debate on rates of depletion may reappear.

OFFSHORE WORKERS' SAFETY AND OTHER RIGHTS

The Department of Energy/Trade and Industry estimated the number of offshore workers in 1979 at 10 500, down from 12 500 the year before (presumably reflecting the large amounts of offshore construction work in the mid-1970s). In 1995, their estimate was 27 300 people working offshore.[6] This is a relatively large population, exposed to considerable process risks which would be unusual even in a normal environment. In the hostile offshore world the risks are doubly serious, and new problems appear, such as transport, navigation and extreme weather conditions.

The UK North Sea has a history of accidents – from the "Sea Gem" incident in 1965 onwards. Some have been caused by the extreme weather; some by carelessness or bad practice; many by helicopters. The worst and most infamous was the "Piper Alpha" explosion in 1988, which took 167 lives.

Piper Alpha, which is described in more detail in the next chapter, led to a new emphasis on safety. When North Sea activity began, the expertise needed to write and administer the rules was in short supply. The management of offshore safety originally fell between several British government departments. The Robens Committee in 1974 judged that "the first and most fundamental [defect] was that there was too much law. This had the unfortunate effect of conditioning people to think of safety and health at work as in the first and most important instance a matter of detailed rules imposed by external agencies".[7] A series of Statutory Instruments since 1972 had introduced regulations on reporting deaths offshore (SI 1542 of 1972); empowering safety inspectors to visit platforms (SI 1842 of 1973); requiring installations to have a Certificate of Fitness approving their design, construction and operating manuals (SI 289 of 1974); requiring life-saving appliances (SI 486 of 1977); requiring fire-fighting equipment (SI 611 of 1978); and so on.

Overall responsibility for ensuring the safety of the offshore workforce was assigned to the Health and Safety Commission (HSC); but in view of its limited offshore expertise it entered, in 1978, into an agency agreement with the Department of Energy, under which the latter exercised this responsibility on HSC's behalf. In 1978 the Burgoyne Committee examined safety again, recommending that the Department of Energy should be given the responsibility of setting

Plate 1 Fire practice taking place on the Tartan platform (Photograph by courtesy of Texaco Ltd)

standards for offshore safety and of ensuring their achievement, using an inspectorate "consisting of well qualified and industrially experienced individuals".[8] Two committee members dissented: they felt that the department responsible for policy (in effect for the rapid exploitation of the offshore resources) could not also be responsible for safety. A new agency agreement in 1981 left the Safety Directorate of the Department of Energy's Petroleum Engineering Division effectively responsible to the HSC for enforcing the 1974 Health and Safety at Work Act (HSWA) offshore, and for advising the HSC on offshore safety standards. The Health and Safety Executive, the executive arm of the HSC, had no specialist offshore capability. The HSC was also advised by the Oil Industries Advisory Committee (OIAC) which included trade union and employer representatives.

Lord Cullen, in his report on the Piper Alpha disaster, was critical of the Department of Energy's attitude to safety: he found its Safety Directorate too small to handle the complex issues properly. Piper Alpha had been inspected by the Department of Energy only a few months before the explosion; Lord Cullen judged that the inspector had "relative inexperience" and was given only "limited guidance". In the early 1990s, the government implemented Lord Cullen's report and transferred control of offshore safety to a new Offshore Division of the Health and Safety Executive (HSE), where it remains at the time of writing. The HSE built up the number of offshore inspectors from a handful to several hundred. It developed teams with specialist expertise, e.g. in diving or drilling operations, who were able to back up the general inspectors.

Lord Cullen also recommended that each installation should in future have a Safety Case, which would be drawn up by the operator and submitted to the HSE. The Safety Case covers the anticipated risks and what is being done to reduce them to acceptable levels; it also covers safety equipment, facilities for emergency evacuation, and so on. The system allows the operator to recommend the best way of dealing with each hazard, the best types of safety equipment, and so on. The HSE then reviews the recommendations. If it accepts them, the Safety Case becomes the basis for future inspections. If not, the company is obliged to rethink and resubmit the safety case. If necessary, the HSE has the right to close down the operation at any time, or to impose conditions.

An important part of the Safety Case is the operator's safety management system. This includes contacts with the workforce: training, drills, and providing opportunities for personnel to raise safety issues and make recommendations. It also includes provision for following up reported defects.

Once the work covered by a Safety Case begins, HSE inspectors visit the operation at random intervals to check that the precautions laid down are in fact being observed. Drafting the Safety Cases was expensive and led to a boom in consultancy work. Reading and assessing them, and making the number of offshore inspections required to monitor them, was also expensive. Safety management involved not only the design but also the operation of platforms, and the onshore response of companies to incidents. The offshore emergency response exercise became a regular feature of industry life, as did training in offshore survival. Almost everyone who goes offshore nowadays has first been turned upside down under water in a specially designed pool, strapped into a mock-up of a helicopter cabin, and required to splutter his or her way to the surface. Visitors to platforms have formal safety inductions: they are shown all the safety equipment, and given a tour of the platform so that they have some idea how to escape from it if the worst happens. New safety features have been fitted. Lifeboats, for instance, are now totally enclosed and able to survive being launched into a pool of burning oil. Some are able to free-fall from the platform into the water to minimise the risks and delays of normal launching procedures.

Trade unions have taken a leading role in pressing for safety improvements, but have been hampered by their lack of membership offshore. This is partly for historical reasons, and partly due to practical problems.

Industrial relations is an area about which there are few hard facts, and many allegations. It seems clear that the unions were slow to recruit offshore in the earliest days of the North Sea, possibly because they assumed that their existing membership in the large oil and construction companies would guarantee them a foothold, or perhaps because they did not anticipate the scale and duration of employment that the North Sea would provide. Once the offshore industry began to get under way, and especially as work moved from construction to operation, the unions discovered practical difficulties that they had not anticipated.

Many offshore workers, unionised or not, are highly nomadic. Although a platform is operated by a particular company and is referred to as, say, a Shell or a BP platform, the workers on it have a range of employers. Caterers, providers of specialist well-logging services, experts on communications, divers, scaffolders and painters, are all brought offshore as required, some for long periods, others for short visits. They move from platform to platform.

Individual companies supplying offshore services may have personnel on platforms all over the North Sea (indeed, all over the world) at any given moment. The terms of service vary greatly. Usually workers spend several days offshore, followed by several days onshore; some commute great distances, e.g. from Spain or the USA. Others live locally, say in Aberdeen, and their work pattern may involve regular short trips to a range of offshore installations. Many are on contract, or work as subcontractors from their own small company. The offshore ethos, too, is result oriented, and increasingly paid by results; there is little tolerance of individuals who do not "pull their weight". Particularly in recent years, unemployment has been rising in the UK, and workers are anxious to retain their jobs. Last, the workforce is international, especially at the specialist level. None of this helps to create a sense of solidarity or a trade union culture.

It is impossible for anyone, including a trade union organiser, to visit an offshore installation without the operator's knowledge and consent, so organising and recruiting members in the workplace is difficult. It is not possible to stand "outside the factory gate". Offshore workers, even if they are union members, come from all over the UK, and may belong to many different local branches of several different unions. Union records are organised by local branches; there is no central means of compiling a nationwide list of those members who are offshore at any given time, or even of knowing at branch level who and where they are, unless they report their own movements to their branches. Identifying non-members and recruiting them is similarly difficult.

Prior to 1979, Labour governments tried to use the licensing process to press the companies to accept unionisation offshore. After 1979, the Thatcher government took active steps to limit union powers throughout British industry. These included stricter rules on balloting, which were particularly difficult to fulfil offshore because of the difficulty in identifying members.

As a result union penetration has been very patchy. Offshore catering workers are possibly the most highly unionised (estimates are about 95%) and their conditions are regulated by an agreement between the unions and the UK Offshore Operators' Association (UKOOA). Many helicopter pilots are members of BALPA, the airline pilots' union. Stand-by vessel crews have a limited degree of union membership. At the opposite end of the scale, union membership on board many drilling rigs (caterers excepted) is probably nil, and is neither recognised nor encouraged by many drilling companies. (Drilling rigs are highly mobile, and their owners are not UKOOA members and therefore not directly bound by UKOOA agreements with the unions.)

The main unions involved are the TGWU, RMT, GMB, MSF and AEEU. Some claim to have up to 6000 members offshore, but union sources suggest privately that the total number of unionised workers offshore is unlikely to be more than 10 000 (or one-third of the number of persons working offshore). It has been suggested that from time to time unions have succeeded in building up significant representation on individual platforms, but this is an area where both sides have an interest in presenting the figures to suit themselves.[9] Strike action has been taken by non-platform workers (e.g. by NUS (National Union of Seamen) members on supply vessels in 1982) or discrete groups of workers (e.g. catering workers in 1980).

The much publicised Oil Industry Liaison Committee (OILC) is not technically a trade union; it was set up as a charity to avoid some of the tighter provisions of anti-union legislation brought in by the Conservative government. Although initially set up and funded by the unions, it has broken from them over questions of tactics and funding. For instance, the OILC called a strike and offshore sit-in in 1990; some 40 installations were affected, but the strain on those "sitting in" offshore became too great, the strike collapsed, and an estimated 1000 jobs were lost.

The unions have a limited number of guaranteed rights, and several agreements with UKOOA. They sit on the Oil Industries Advisory Committee (OIAC) which advises the Health and Safety Commission. Representatives of the Inter Union Offshore Oil Committee have to be allowed to visit offshore installations or rigs twice a year, but this is at a time and place chosen by the oil companies. Labour Party spokespeople have said that they believe that offshore union representation is essential; this, if it came about, would presumably be part of a wider package involving the repeal of some present industrial relations legislation and the introduction of some provisions based on the European Social Charter.

Industrial relations remains a sensitive area. Some union leaders privately say that most large companies in the industry, while not welcoming, are responsible and relatively easy to deal with. Problems come particularly with some smaller companies and with contract staff, who have only limited rights of recourse to an industrial tribunal. Foreign workers are allegedly brought in at cheap rates, by some companies, to fulfil some roles offshore (e.g. painters or sandblasters). Official statistics suggest that foreign workers are a very small percentage of those employed offshore, but this may be open to interpretation. The OILC has also made allegations of victimisation of offshore workers who try to promote trade union activity, or to report safety issues. There is now provision for workers to report safety issues confidentially to the HSE; this is little used, which may suggest either that fears of victimisation are genuine, or more probably that there are already enough channels for such reports to be made.

"Offshore" remains a largely masculine world, although the numbers of women offshore are increasing slowly. Equal pay and rights were extended to offshore workers by a Statutory Instrument in 1987 (SI no. 930.) Onshore workforces, too, are largely male, and "glass ceilings" appear to exist in several companies – although the head of one UK operating company, Oryx Energy UK, is female.

JOBS FOR BRITAIN

An early government objective was to secure work for British industry. Originally it had been intended to make BP, in which the British government had a large shareholding, the major player in the North Sea. (This was partly achieved: in 1994, BP produced 17% of UKCS crude oil and 13% of UKCS gas.)

Some work would inevitably come to the UK. Constructing the major platforms and pipelines had to be done as near to the fields as possible. New yards were built in Scotland to take them: for instance, a dock at Nigg for the construction of the Forties platforms, or the £26 million, 3.5 acre drydock built at Kishorn in 1975, to construct the enormous concrete base of the Ninian Central platform. UK shipyards were less able to compete in the market for mobile drilling units (MODUs) or "rigs"; these have been built in areas as diverse as Singapore, the USA, Japan and France.

Plate 2 The jacket section of a production platform for Forties is winched out of Nigg Bay in August 1974 (Photograph courtesy of British Petroleum)

The offshore platforms required a growing amount of services. These included maintenance and such things as painting, electrical and communications equipment, catering and the supply of new videos for onboard entertainment. They needed marine experts to plan and supervise their moves, tugs to tow them and lay out their anchors, heavy lift cranes to install and modify them, specialised supply vessels ferrying a constant stream of goods out to them, and stand-by vessels to maintain a lonely and dogged watch, required at all times by law to be within five miles of the installation in all weathers ready to pick up anyone fallen overboard. They needed helicopters, the bus services of the North Sea, constantly changing the platform crews over. Aberdeen airport became the busiest helicopter airport in Europe. Some services were supplied by indigenous companies, such as the Wood Group in Aberdeen. Others were supplied by external companies.

In 1979[10] the government estimated total capital investment in oil and gas at £2 billion – about 6% of total UK investment. No figures were given for the share of this work done by British industry. In 1984, annual investment in oil and gas had reached £3.2 billion: 24%

of total UK industrial investment. The total value of orders reported by operators was £3.6 billion, of which 74% or £2.65 billion was "the UK share". By 1989, gross capital investment in the oil and gas industries was £2.6 billion, representing "about one eighth of total UK industrial investment". The total value of orders reported was £3.9 billion, of which the UK share was 81% or £3.2 billion.

These government figures have been questioned. Much of the investment undoubtedly goes to foreign firms operating in Britain, and though these provide employment and pay taxes it is arguable how they should be counted in the statistics. Whether "British industry" could have made more of the opportunities is a wide issue, going beyond the scope of this book and requiring an assessment of Britain's post-war society and industrial competitiveness as a whole. This is perhaps best left to a future generation of writers; it is still difficult to be objective.

Some of the arguments have already been set out. Christopher Harvie's fascinating history of the North Sea industry, *Fool's Gold*,[11] makes a case that poor planning and lack of government action minimised the possible returns to Scotland. Evidence that there was at least some benefit comes from Grampian Regional Council (which was responsible for the Aberdeen area until 1994). The Council has identified four benefits of the oil boom for the region. The first was 40 000 to 50 000 jobs directly created by the industry, plus others indirectly due to it. Locally based firms which were "100% involved in the oil industry" employed an estimated 20 000 people in 1976, and 51 500 in 1993. Average earning levels increased: in 1972 the average weekly earnings of a man in the region were 88% of average UK male earnings; in 1992 they were 111%. (Women's earnings also increased, but by less, as might be expected in a predominantly male industry: they were 87% of the national average in 1972, and 94% in 1992.) The region had better access and services, and its entrepreneurial spirit had been encouraged. Looking at the disadvantages, the Council noted that the pressure on wage rates had damaged traditional local industries. The region had lost its access to UK and EC development grants, as it was no longer poor enough to qualify. Housing had become scarce and prices had risen. (In 1972, the average price of a house in the region was 73% of the UK average. In 1992, it was 98%. In 1982, at the height of the boom, it had been 126%). Last, the council itself had been obliged to spend over £100 million on oil industry related infrastructure, in some years as much as 12% of its budget.[12]

In 1973 the government set up the Offshore Supplies Office (OSO), a branch of the Department of Trade and Industry, to ensure that British suppliers were given the best opportunities to market their goods and services in the UK and overseas. British specialised expertise and products became an international export. One of OSO's directors-general, Norman Smith,[13] was later to say that Britain's worst record was in marine activities requiring large capital investments – drilling rigs, diving support vessels, heavy lift and pipe-laying vessels, for instance. He laid the blame partly on the lack of risk capital from the City, and partly from the inexperience and undercapitalisation of some British companies. (OSO has since been renamed the Oil and Gas Projects and Supplies Office, but has retained the "OSO" logo.)

Early attempts to introduce codes of practice, or to set operating licence conditions, in order to persuade the operators to buy British had only qualified success. (After all, what is British? Sometimes it would take an accountant to pull apart every component of a complex product, decide where each had come from and where the profits of making it had gone, and produce a balance sheet.) As Britain became part of the European Union, such codes became increasingly impossible to maintain. EU directives now set serious limitations on any attempt to apply "buy British" rules to the offshore industry (see Chapter 9). If there ever was an industrial El Dorado beyond the opportunities which British industry has already found and won, then it has probably vanished for ever.

THE ENVIRONMENT

Historians have distinguished two environmental "waves". The first, starting in the 1960s and most famously expressed in the 1972 Club of Rome report on the limits of growth, focused largely on the resources of the planet as a whole and the need to keep consumption and replacement in balance, particularly of such things as fossil fuels. The second wave, starting in the late 1970s or early 1980s, tended to look at smaller specific issues; it was also much more politicised, either through normal party mechanisms or through protest action of one sort or another. The UK Green Party won 15% of the UK vote in the 1987 European elections (although this did not entitle it to any seats). Since then the political fortunes of the Green movement have waned;

but its influence on government (exerted directly, via pressure groups or public opinion, or through the European Union where Green political parties have a stonger foothold) has had considerable impact on industry and others, in the form of new and tighter regulations.

Until 1995, the UK offshore oil industry had never been seen as a major environmental villain. If the North Sea was mentioned in contemporary environmental literature, this was almost always in connection with the dumping of chemical and other wastes, not connected with the offshore industry.[14] There are several reasons for this. The offshore oil industry is out of sight and out of mind; and the amount of pollution it causes is small compared with other sources. There were more serious concerns: as an example, North Sea related Greenpeace actions before 1994 focused on the disposal of large amounts of industrial waste into the North Sea by Tioxide and other European titanium dioxide producers (1984), the carriage of uranium hexafluoride by sea from Belgium to Riga (1984); further titanium oxide dumping (1985); UK sewage sludge dumping (1987); the deliberate incineration at sea of toxic hazardous waste such as PCBs (1987); toxic waste dumping by ships (1988); and alleged British government inaction over the outbreak of "seal plague" (1988). The Ekofisk Bravo blow-out, which took place in the Norwegian sector of the North Sea in 1977 and spilled 9000 tonnes of crude, has been accurately described in a 1991 textbook on the environment[15] as "less an environmental disaster than it could have been, as the wind kept the slick away from the shorelines, allowing it to be broken down by wave action and chemical dispersants". The same book points out that only about 5% of oil reaching the ocean comes from such accidents as blow-outs or tanker disasters.

Offshore platforms are generally well regulated; they cannot dump wastes over the side in the way that unscrupulous ships' captains may permit, if only because they cannot sail away from the evidence. Oil refineries and terminals onshore are also carefully operated, but they are undeniably ugly, locally intrusive, and prone to emitting foul smells however responsibly they are managed. When the oil and gas industry suffers from a poor environmental image, it is usually due to onshore activities or to major oil spills from tankers.

For the purposes of this book, however, it is important to distinguish the offshore industry from its refining and marketing colleagues, or from the more general issue of "the limits to growth".

There are three specific environmental problems caused by offshore production.

The Risk of Oil Spills

Major oil spills have the most spectacular impact of all pollution, but they are of the least relevance to the UK offshore oil and gas industry. Offshore installations (as compared to oil tankers) have not in the past posed a threat to the UK coastline: the majority of them are simply too far away for the small amounts of oil that have been spilled to reach the coasts. (However, with the development of fields in areas closer to shore, such as Morecambe or Weymouth Bays, any spillages will become more sensitive.)

Accidental oil spills from offshore installations are, of course, something which all operators strive to prevent. They are expensive in terms of lost oil, as well as regards any possible environmental consequences. They usually occur as a result of other failures or accidents, and may involve safety risks and damage to installations. The evidence varies. DTI figures for spills from offshore platforms between 1985 and 1994 show some bad years (e.g. in 1986 a total of 3540 tonnes of oil was spilled, and in 1988 a total of 2627 tonnes; but in each case these were largely attributable to one or two major spills: in 1986 one of 3000 tonnes, and in 1988 one of 1500 and another of 750 tonnes). However, from 1991 to 1994 the figure has varied between 225 and 174 tonnes, all of these being relatively small. The only major North Sea blow-out in 30 years was in the Norwegian sector, from Ekofisk Bravo, in 1977. As noted above, this did not cause serious problems onshore.

An offshore oil spill poses a variety of environmental hazards. Crude oil is composed of a wide range of hydrocarbons and other substances, and each crude is different. Light oils, such as many North Sea crudes, evaporate relatively quickly, and many of their components disperse naturally in the water column. Very little of the spill is left on the water surface after 48 hours. Heavier oils remain longer on the surface. In heavy seas, oil on the surface may form a mousse or "emulsion" with water, which takes longer to disperse.

Oil spills out in the North Sea rarely reach the shore; the main hazard they pose is to seabirds, including some species (such as guillemots and razorbills) which moult their flight feathers for part of each year and are forced to spend this time flightless on the water surface. The consequences of even a small spill in the moult area of a colony of these birds would be very serious, since they would be unable to get out of its way. Much depends on where and when the spill occurs. According to the Nature Conservancy Council,[16] the Ekofisk blow-out in 1977 caused very few seabird casualties. Smaller spills, particularly closer inshore, may have catastrophic consequences, but the Council judged in 1987 that: "Undoubtedly oil has been spilt as a result of offshore production in the North Sea, but to date no damage to bird populations caused by oil from the platforms has been demonstrated."[17]

To help protect seabirds, the Nature Conservancy Council set up in 1979 a Seabirds at Sea team to monitor seabirds and to assess where birds are particularly vulnerable to oil or other pollution. This information is used in contingency planning and is also a factor in the environmental assessment of licensing. It helps the UK to fulfil such international obligations as the European Union Directive on the Conservation of Wild Birds (1979). The Seabirds at Sea Team produced reports detailing the location and vulnerability of birds in the North Sea (1987), and to the west of Britain (1990), to the south and south-west of Britain (1995), and also an atlas of seabird distributions in NW European waters (1995).

Oil wells closer to the shore require much more detailed environmental approval. If there is a possibility of oil spill reaching the shore, or, in shallower water, of damage to fish or sea-bed creatures, the Department of Trade and Industry applies more stringent licensing rules. Environmental impact assessments are made and lodged with the Department. A pollution response plan has to be prepared by the operator and agreed; this will typically involve having one or more pollution control vessels on stand-by. The use of dispersant on spilled oil has to be carefully considered: dispersant may do more harm than good, and its use is subject to licence by the Ministry of Agriculture, Fisheries and Food (or its Scottish equivalent, the Scottish Office Agriculture and Fisheries Department).

The majority of oil spills in the North Sea have not come from oil exploration and production activities; the larger ones have come from oil in transit through the area on tankers (e.g. the *Braer* spill in 1994). The British government's response to these spills is coordinated by the Marine Pollution Control Unit (MPCU), originally part of the Department of Transport and now of the Coastguard Agency, and by the local authorities whose shoreline is affected. Although the treatment of oil spills at sea and on the shore is a controversial subject, offshore industry spills have not so far been of any seriousness and the issues are not directly relevant to this book.

Non-accidental Emissions into the Sea

The second of the environmental problems potentially caused by the offshore industry includes the dumping of "mud", cuttings from the wells, and water produced from the wells, or other wastes, into the sea. The drilling "mud" used in wells is often an oil-based solution. It also contains some toxic elements. However, the percentage of wells using oil-based mud has declined (in 1985, 211 out of 290; in 1994, 102 out of 309[18]) partly in response to stricter environmental regulations; the level of oil which may be discharged with cuttings is now 1% for new exploration and appraisal wells. Cuttings – i.e. the rock extracted as the well is bored – are cleaned to remove mud before being dumped. Water produced from wells is also separated from the oil and gas, and then cleaned before being discharged back into the sea. The amount of oil discharged with produced water has risen over the last 10 years (in 1985, 2150 tonnes; in 1994, 4418 tonnes) as has the number of installations permitted to discharge it (in 1985, 34; in 1994, 52). This is largely because, as reservoirs age, more water is produced with the oil; as a percentage of total water discharged, discharged oil is not allowed to exceed 40 parts per million by weight.[19]

All discharges are covered by the Prevention of Oil Pollution Act, 1971, which requires operators to seek permission before dumping any oil-contaminated cuttings or water. Maximum levels of oil permitted in these dumpings are set at levels determined by the Paris Commission.

Emissions into the Atmosphere

Atmospheric emissions are the third area where the offshore industry has a potential impact on the environment. There has been considerable interest in atmospheric pollution over the last few years. Acid rain and the "greenhouse effect" have made all countries more conscious of what is released into the air.

There has been some disagreement over the figures produced by various investigations into atmospheric emissions from offshore. Methane emissions in particular have been the subject of recent argument. For simplicity, and to give a general idea of the scale of releases involved, a survey by an industry body is quoted here. The United Kingdom Offshore Operators' Association (UKOOA)[20] commissioned a study to determine the levels and types of atmospheric emissions resulting from offshore activities. This found that the largest emission in 1991 was carbon dioxide at 19 million tonnes, which is only 3.2% of the UK total. (Domestic and commercial users produce 20.8% and road users 19.8%.) The main sources of carbon dioxide emissions offshore are power use and flaring of produced natural gas. Power use offshore is essential, as in onshore processes. (Some platforms produce enough electricity for a small town.) Flaring has been drastically reduced in recent years, partly by government action and partly by the realisation that flaring gas is, almost literally, burning up money. Government consent is required, except in emergencies.

Emissions of methane from offshore facilities, at 99 000 tonnes, were only 2.9% of the national figure. The bodily functions of cattle accounted for 23% of the national total, and those of sheep for 11%. Methane release can occur when pipelines or production facilities are vented for maintenance or safety reasons; the UKOOA study traced 30% of the upstream oil and gas industry's emissions in 1991 to four specific venting operations. The industry findings have been challenged by the Watt Committee on energy and a Department of the Environment study,[21] but it still seems clear that the offshore industry is a relatively minor source of methane.

According to UKOOA, the industry accounts for 4% (127 000 tonnes) of the UK's nitrous oxides (NO_x) emissions, almost half of this from power generation offshore and at terminals; and 2.8% or 74 000 tonnes of volatile organic compound (VOC) emissions. According to UKOOA, the industry produces 67% of UK primary energy but emits only 3% of total UK global warming gases.

Abandonment of Offshore Platforms

The most controversial environmental issue at the time of writing is undoubtedly the abandonment of the platforms. The furore which, in

1995, surrounded and succeeded in reversing Shell's decision to dump the Brent Spar platform in the deep Atlantic shows that this is an issue for the future. It is covered more fully in Chapters 4 and 11.

REFERENCES

1. A Jones, *Oil, the Missed Opportunity*, Andre Deutsch, 1981, p 155.
2. KPMG Peat Marwick, *Oil and Gas: a Survey of Published Accounts*, 1990, p 4.
3. See S Brittan and B Riley, *A Peoples' Stake in North Sea Oil*, Unserville State Papers no. 26, Liberal Party Publications Department (no date, but the proposal appears to have first been made in 1978).
4. A Jones, *Oil: the Missed Opportunity*, p 165.
5. House of Commons Select Committee on Energy, 3rd Report, May 1982, p xliii.
6. Department of Energy/DTI "Brown Books" for 1980 (p 2) and 1995 (p vii).
7. Quoted in Cullen Report, HMSO Cmnd 1310, 1990, p 256.
8. Ibid., p 258.
9. See S S Andersen, *British and Norwegian Offshore Industrial Relations*, Avebury/Gower, 1988. Andersen quotes no authority for his statements that, e.g. "The membership rate on Shell platforms was in 1985 about 40%" or [in 1984] "ASTMS had about 70% membership on Brent Bravo".
10. All figures from Department of Energy/DTI "Brown Books" for the year after the one cited.
11. C Harvie, *Fool's Gold*, Hamish Hamilton, 1994.
12. Figures from Grampian Regional Council Economic Development and Planning Unit, *Oil and Grampian – the First 25 Years*; no date, but around 1994.
13. Quoted in C Harvie, *Fool's Gold*, p 222.
14. See, for instance, the Open University textbook, P M Smith and K Warr (eds), *Global Environmental Issues*, Hodder and Stoughton, 1991; M Brown and J May, *The Greenpeace Story*, Dorling Kindersley, 1989; E Goldsmith and N Hildyard (eds), *Earth Report 2*, Mitchell Beazley, 1990.
15. Smith and Warr, *Global Environmental Issues*, p 63.
16. Nature Conservancy Council, *Vulnerable Concentrations of Birds in the North Sea*, 1987.
17. Ibid., p 7.
18. Department of Trade and Industry "Brown Book", 1995, p 26.
19. UK DTI Minister, Lord Fraser of Carmyllie, quoted in *Petroleum Review*, December 1995, p 552.
20. Dr B G S Taylor, UKOOA, article in *Petroleum Review*, April 1994, p 174.
21. See *Energy Policy*, 1995, vol 23 no 3 p 244.

3
The Nerve-wracking 1970s, the Booming 1980s

Seven British governments have grappled with the political problems described in the previous chapter. The first round of offshore licensing had taken place in 1964, under Sir Alec Douglas-Home's Conservative government; 53 licences were given to 51 companies. The round, according to Aubrey Jones, "sought to secure two aims: first, to favour British companies like BP; secondly, to accelerate exploitation".[1] The then Minister of Power, Frederick Errol, later told the *Financial Times*: "I rushed these through before the General Election ... I was afraid that a socialist government would get in and refuse licenses to private enterprise."[2]

Despite political uncertainty in the 1970s, as the Conservative government of Mr Heath struggled with the trade unions, the oil companies had continued to explore. They began to make major finds at the beginning of the 1970s. Finding a giant field is only the first step down a long and expensive road: committing the capital to follow that road until the field produces a positive cashflow can be a serious, "bet the company", decision. In 1970, the price of oil was so low that it was not easy to justify the enormous expenditures involved; the giant oil fields were in the northern and central North Sea, far deeper and more hostile waters than the southern areas where the first large gas fields had already been in production for some years.

Since the end of the Second World War, the price of crude oil had stayed low: in mid-1973 it had reached $2.59 a barrel. The oil-producing countries were aware that they raised less in tax from each barrel than governments in the consuming countries, who imposed high taxes on gasoline and other uses of oil. Resentment built up and led to the formation of OPEC in 1960, but even OPEC had made little initial impact on the price.

On 6 October 1973, however, Egyptian forces invaded Israel. This catalysed a reaction that had been building up for many years. The Arab producers began to limit oil supplies and raise prices, partly to dissuade western consuming countries from supporting Israel, and partly to increase their own revenues. OPEC's preparations paid off: the infrastructure was in place for a coordinated effort by the producing countries. Within three months the oil price had risen fourfold, and it had once again become clear that oil was a strategic commodity. The spectres raised in the 1950s by Iranian nationalisation and Suez returned; the oil-producing countries (usually thought of in Britain as "the Arabs") could hold the consumers to ransom. No one could see an end to the process; no one could dare to suggest an upper level which the oil price might not pass.

Of course, North Sea oil also quadrupled in value. The high extraction costs suddenly became much more realistic. Oil under UK control offered security of supply not only to Britain itself, but also to the international oil companies, which depended on a continuous supply of crude to feed their extensive refining and marketing networks.

THE DREAM OF NATIONAL WEALTH

Harold Wilson's first Labour government, which took office in 1964, may or may not have been the socialist one that Frederick Errol dreaded. It was replaced in 1970 by Edward Heath's Conservative government. As the 1970s began, the greatest political concern for Conservative ministers and oil companies alike was left-wing enthusiasts like Tony Benn, the Minister of Technology from 1966–70 in the first Wilson Labour government, who did finally become Energy Minister in 1975 under Mr Wilson's second government. A peer who had renounced his title and identified increasingly with the left wing of his party, he was regarded with suspicion by political enemies. Echoing Aubrey Jones's fears that the oil might run out, he argued that it should be used for specific programmes to revitalise the economy.

> As a nation, we are now living on the unexpected, and strictly temporary, legacy of the oil revenues. Instead of using that money to re-equip our industrial base and to sustain our economy and living standards when the oil runs out, we are now using it to finance tax cuts for the wealthiest section of the community . . . Unless the present policy is changed . . . when the oil reserves have been exhausted, we shall be left with such a narrow industrial base that it will not be strong enough to sustain our population.[3]

The Labour government also intended to intervene much more directly in the production process. They were not alone in this; many other oil-producing states were suspicious of the companies and the power they wielded.

It is from this period that the exploitation of Britain's oil resources really begins. The first offshore oil field to produce was Argyll, in 1975; the Hamilton Oil company got a head start and avoided the costs of building a fixed platform, by using a semi-submersible rig as a floating production facility. This was a world first,[4] although it is now increasingly common – see Chapter 6. Later in the year the Brent field began to produce.

More northerly fields required larger investment, larger offshore structures, extensive pipeline systems to bring back the oil, and new processing plants to handle it onshore. Development work on the Ninian field began in earnest in 1974; the first oil did not come until 1978. The scale of these deepwater fields was enormous: the construction costs of the three Ninian platforms and associated pipelines were estimated at £1.2 billion.

At least one oil company, Burmah Oil, caught up with the enthusiasm, overstretched itself to the extent that it was unable to meet its borrowing commitments, which had to be guaranteed by the Bank of England. (Denis Thatcher, husband of the future Prime Minister, was one of Burmah's directors.) Burmah had gone into refining and liquefied natural gas (LNG) transport, and also owned several North Sea blocks. Its collapse was an illustration that oil companies do not always know best, and must have encouraged advocates of state control.

As the massive sums began to be disbursed, the government passed the Petroleum and Submarine Pipelines Act of 1975. This was the much-feared attempt to change the nautre of the UK offshore

Plate 1 The first North Sea oil flows onshore in June 1975, as Energy Secretary Tony Benn and Fred Hamilton of Hamilton Oil turn on the taps (Photograph courtesy of BHP Petroleum Ltd)

industry, bringing it more under government control. A British National Oil Corporation (BNOC) was created, to take shares (on a commercial basis) in licences, and to trade in oil on behalf of the government. It was given a ragbag of assets and powers, stakes in some new fields, and the offshore assets of the Burmah Oil company.

Many national oil companies were set up in the mid-1970s – for instance the Venezuelan state company PDVSA – and BNOC was not untypical. It was a state oil company which many on the left hoped was the first step to total state control of the North Sea. (BP, although half-owned by the British government, operated as an independent commercial company.) The government had, however, stopped short of outright nationalisation of existing licenses. The Petroleum and Submarine Pipelines Act contained some provisions retrospectively altering the terms of existing licences, such as new ministerial powers over gas flaring, depletion rates and work programmes. The

Conservative opposition argued that, by reducing the value of existing licences, the government was practising expropriation without compensation; but the "expropriation" had only minor effects. The question was whether it was to be the thin end of a wedge.

In the 1977–8 and 1978–9 licensing rounds (numbers 5 and 6) the British government insisted that 51% of the crude oil produced must be sold to BNOC. BNOC's other functions were to provide expert advice to the government and to extend government control over exploration, development and production. It could use its powers to discourage companies from tying up British oil in long-term contracts to foreign customers, which might deny supplies to Britain in any future crisis. It could purchase shares, on a commercial basis, in operating partnerships, but it might then use its vote as a shareholder to promote government policies rather than the best economic interests of the partnership. Presumably to placate the Scots, BNOC was housed (like the OSO) in Glasgow, where it was inconvenient for both London and Aberdeen.

Another issue, which loomed larger at the time than subsequent events have yet justified, was the possibility of devolution for Scotland. The Scottish National Party (SNP) had a third of the popular vote in Scotland in 1974, and its main policy platform was secession from the UK. As a separate country, and assuming the Shetland Islands were included, Scotland would in theory have controlled many of the UK's offshore assets: all the northern and most of the central North Sea fields, including the giants such as Forties, Ninian and Brent. The economic policies of any Scottish government were an unknown factor. Scottish nationalism remained a live, if to many English policy makers unthinkable, issue until a referendum in March 1979 and the fall of the Callaghan government effectively put an end to the risk of secession – at least for a time.[5]

These were the years in which the first serious attempt was made by a British government to work out the control of the North Sea; but it was made against a background of continuing crisis. Britain did not feel rich. The national champions were the Labour governments of Wilson and Callaghan, with successive Energy Ministers Eric Varley and Tony Benn. The role of North Sea oil was clearly going to be crucial to the economy and the survival of governments. It was also caught up in the internal struggles within the Labour Party, between the pragmatists (such as Wilson and Callaghan) and those like Benn who believed in applying socialist ideas as far as possible. The offshore issues were highly technical; actual industry operations were remote and affected relatively few people; and the general political debate seems to have focused more on how to spend the money that the North Sea would produce than on how the production should be regulated or how the profits should be shared between industry and government. The growing power of the trade unions was perhaps the major economic issue; British industrial relations were confrontational and led to strikes and economic inefficiencies – as the major oil companies knew from the restrictive practices with which they had to contend in their refineries.

Meanwhile, the "energy crisis" was in full swing, with restrictions on the use of electricity and other measures to cut down oil and coal consumption. The 1976 Energy Act, introduced "to make further provision with respect to the nation's resources of energy", was a crisis measure designed to minimise energy consumption and maximise production.

National politics were dominated by one economic disaster after another. Although Britain now knew it had a major asset in the offshore hydrocarbon resources, relief was not yet at hand for the economy. The international economic system was in chaos; the increased oil price had effectively transferred very large amounts of money from oil-consuming countries (such as Britain) to oil producers. The oil consumers faced a sudden and enormous drain on their national resources. The oil producers could not absorb their windfall, and much was reinvested, causing currency instability and, among other things, an uncontrollable fall in the value of sterling. British foreign reserves were spent in a vain attempt to defend the pound, and in September 1976 the government was again forced to seek long-term International Monetary Fund (IMF) support. As in 1968, this came with painful and humiliating conditions: "it was a period of extraordinary overseas influence on British economic policy", wrote David Smith.[6] While much of the framework for exploitation of the North Sea was being decided, Britain was facing unprecedented inflation rates of over 20%, unemployment was rising, and the government changed back and forth between the two main political parties.

Little wonder that the North Sea was not a policy priority, even though Department of Energy estimates were that Britain might produce up to 3 million barrels a day – well above national requirements. In 1977 the Labour Minister of State for Energy (i.e. junior minister),

Dr Mabon, said: "Within three years Britain will be one of the ten largest oil producers in the world; and one of the very few industrialised countries to be self-sufficient in oil and gas." North Sea gas, although produced for domestic use only and therefore not contributing directly to the balance of payments, was estimated in 1976 to save the country £2.25 billion per annum.

In these circumstances, the oil companies wielded considerable bargaining power against the Wilson and Callaghan Labour governments of 1974–9. The House of Commons Public Accounts Committee had found that the companies had paid almost nothing in tax between 1965 and 1972. A White Paper issued in July 1974 announced the Government's intention to "secure a fairer share of the profits for the nation". Draft oil tax legislation issued in November 1974 provided for a Petroleum Revenue Tax (PRT) to be set at 60%, in addition to corporation tax. The companies responded quickly, entering into detailed negotiations with the government, through a committee of the United Kingdom Offshore Operators' Association (UKOOA).

Much has been written about these negotiations.[7] It seems clear, however, that the government was seriously distracted – it was fighting off demands for Scottish independence – and also that the oil company lobby was effective. The companies argued that too high a tax rate would kill off all development plans for "marginal" fields. The UKOOA response, as described in the press at the time, quoted a list of fields which would be "economically marginal at todays construction costs and with the additional burden of PRT".

As a result of industry representations, the government modified its proposals. The basic PRT rate was to be set at 45%, and there were to be a host of exemptions and "safeguards". For example, the first million tonnes of oil produced from each field each year were to be exempt. (This provision was clearly meant to help the smaller fields. Heather, for instance, cited in the UKOOA document as one of the fields that would become "economically marginal", was developed and produced its first oil in 1979; but only in three years since then (1982–4) has it produced significantly more than 1 million tonnes.) In 1975, total UK oil production was only just over 1 million tonnes per annum; but by 1979 Forties was producing over 24 million tonnes per annum, Piper 13 million, and Brent nearly 9 million. The Oil Tax Act and the Offshore Petroleum and Pipelines Act became law in 1975, setting a framework for government control of the new resource.

Ironically, however, relatively little tax revenue was to flow to the government until after the fall of the Labour government: Mrs Thatcher's Conservative government was the real beneficiary.

Almost as much politicking was taking place among the companies themselves. The 1970s were the period in which the really large investment decisions were taken and the major fields developed. This called for the expenditure of heroically large sums of money, and held out the possibility – but not the certainty – of massive rewards. In shipyards and at remote Scottish sites the major platforms were slowly growing, eating their way through a mountain of cash. Ninian, for instance: its discovery was announced in 1974; Chevron was appointed operator in 1975; the Southern platform was towed out in 1977, the Northern and the gigantic Central platforms in 1978. First oil was produced in 1978, and all three platforms were producing by 1980. In that year, the UK Department of Energy estimated the UK's total recoverable reserves of oil at between 2200 and 4400 million tonnes, of which a mere 179 million had been produced by the end of 1979.

The problem for the oil companies was that not every oil or gas field is "economic". The decision whether a field is worth developing depends on the oil company's expectations of the return versus the costs. This is not a simple sum. The value is a function of expected future revenue from oil or gas sales, less expenditure on development (capex) and maintenance and operations (opex) as well as taxation and other costs. As explained in Chapter 2, these are typically expressed as an internal rate of return (IRR), in order to compare different projects. In assessing projects, companies usually set a target IRR; if the project is not expected to produce the target rate of return or more, it is unlikely to be agreed.

For the companies, raising capital was not easy. Modern innovations, such as project-specific loans, were not available; only the larger integrated "major" oil companies, with constant cashflow from their retail operations, had the resources or the credit-worthiness to undertake such massive projects. From the first, they spread the risk by forming partnerships to develop fields. Arriving at an IRR or an NPV requires a set of judgements, which each company in a partnership will make differently. Future revenue from sales will depend partly on the size of the reservoir and the amount that can be recovered from it, but also on the oil or gas price. Capex and opex are also difficult to estimate, particularly in the earlier years of exploiting

62 Waves of Fortune

Plate 2 Towing out the jacket for a large platform: Texaco's Tartan platform begins its life (Photograph courtesy of Texaco Ltd)

Plate 3 A heavy lift crane barge prepares to lower the first Tartan platform module onto the jacket (Photograph courtesy of Texaco Ltd)

Plate 4 The platform takes shape: the Tartan platform accommodation module is fitted (Photograph courtesy of Texaco Ltd)

a particular part of the world. (Development costs for the Auk field, for instance, which began production in 1976, came in at 260% over budget.[8]) Tax rates change. Even the assumed rate of interest used by different companies will vary: it is at best an assumption. The effect of changes in this assumption can be considerable, since they affect the financing costs. Any one of these factors can turn a field from an attractive proposition into a loss maker. Almost every project is a "marginal" field, or one which is only economic under certain sets of assumptions.

The complexity of decision making multiplies when it is carried out by a committee. Partnerships vary greatly. At one extreme, Shell and Esso share some fields on a 50/50 basis, with Shell as the operator. Other relationships are more complex. Piper, for instance, was licensed to Occidental, Thompson Newspapers, Getty Oil and Allied Chemicals. Decision-making structures within a partnership vary: they are determined by the partners, and not prescribed by UK law. Typically some form of majority voting is required to carry out any decision. (A partner who objects must either go along or sell its share.)

If a field development programme is to be agreed, all partners must find large sums of money. Each company involved will consider the investment it is called to make against several criteria. These will include how the development fits into its own pattern of activities and investment elsewhere. If it is short of crude, it may be prepared to invest more heavily than a company which is crude rich. Much will depend on its own cash position and its ability to borrow. Its tax position (under a PRT regime, for instance) may also influence its decision: some companies will have taxable income which can be offset against expenditure, others may not. Last, even such technical issues as different interest rate assumptions or company IRR targets will lead the partners to evaluate a project differently.

These arcane technicalities have had several effects on the development of the North Sea. First, they have meant that some fields have been developed as a compromise between different company requirements, rather than in the way thought at the time to be most effective. The Piper field, for instance, was developed as a one-platform field. It was argued[9] that two platforms would have maximised the revenue from the field, but that the higher cost was too much for some of the partners, and a one-platform approach was adopted as a compromise.

A second effect of the differences in company approach to economic analysis has been that some operators have been prepared to take on fields which others would not. (Unocal's Heather field, for instance, was widely regarded at first as a very marginal project.)

In 1979, the Labour government of James Callaghan was coming to an end. A second wave of dramatic oil price rises had begun in 1978, with the collapse of the Shah's government in Iran and the abrupt loss of Iranian oil supplies. This led to the collapse of many long-term contractual agreements and frantic trading in the spot markets, bidding up prices to over $30 per barrel in 1979. The government announced an increase in PRT rates from 45% to 60% and a halving of the tax-free amount of 1 million tonnes per platform per year, in order to capture the windfall profits now being made from North Sea oil. The Chairman of BNOC was also accusing the companies of sitting on licence blocks granted in earlier rounds and refusing to develop them; and by the time of the Sixth licence round in 1978/9, licence criteria were being toughened to increase the powers of BNOC. It was becoming very difficult to assess the true costs and returns of the fields: no one knew what the oil price would do next.

The industry was now capturing the public imagination and holding out a new hope of economic security. In 1978 the UK even issued two postage stamps, as part of series on energy, showing an oil production platform and a natural gas flame.[10] By 1979, total offshore production of crude oil was 78 million tonnes, up from 1 million in 1975; consumption was 94.5 million. Gas production was relatively unchanged at 34.2 mtoe, with consumption at 41.9 mtoe. The major fields were on stream. Forties began production in 1975, Brent and Piper in 1976, Ninian in 1979. Indeed, one or two fields, such as Auk which came on stream in 1976, had already in 1979 passed their peak production and begun to decline.

THE THATCHER YEARS AND AFTER

The polarisation of British politics led to an equally radical swing in 1979 when Margaret Thatcher's Conservative government was elected to office. The daughter of a Grantham grocer, Mrs Thatcher headed a party traditionally led by the sons of aristocrats. As iconoclastic in her own way as Tony Benn, she disliked almost every kind of government intervention, and although she lost office in 1990 her views have dominated British politics ever since. (Conservative Party policy on the energy industries was set out in 1995 by the Minister of State for Energy, Tim Eggar, in terms with which Mrs Thatcher would surely agree: "The apparently disparate objectives of national competitiveness and sustainable development do ultimately have a common theme – they are both essentially about making the best possible use of resources. The Government's view is that the optimal use of resources can for the most part be left to competitive forces in free markets."[11])

Inevitably, BNOC was disbanded. As Aubrey Jones wrote in 1981, "the United Kingdom is unique, having begun to travel along the same route as the United Arab Emirates and Canada in establishing a state company, and then having gone into retreat by curtailing the

Corporation's activities and powers."[12] The 1982 Oil and Gas (Enterprise) Act hived off its offshore interests to a new company, Britoil plc. (BNOC was left with the main function of acting as an oil trading company. With the change in the markets, after the Iranian revolution, from long-term to spot trading, it was felt that this activity was too risky and too commercial to be carried out by a government body, and the 1985 Oil and Pipelines Act abolished the Corporation altogether.) The British Gas Corporation, which had held far tighter control over Britain's gas than BNOC ever held over oil, was privatised (as British Gas plc) in 1986.

In 1987, the government sold off the controlling share it held in BP – just one of many sales of national assets that introduced the word "privatisation" to the economic vocabulary. (The government's tolerance of the free market was not limitless; when the Kuwait government snapped up 22% of BP, it was forced to reduce its holdings to a more acceptable 10%. After all, it was only in 1975 that Kuwait had cancelled BP's concession to produce Kuwaiti oil, offering only small compensation.)

Tax policies were less welcome to the companies. The rate of PRT was quickly raised to 75%. The tax take was at its highest between 1981 and 1986, and the Conservatives introduced an additional Supplementary Petroleum Duty in 1981 and 1982. Much later, in 1993, the Conservatives were to remove the PRT altogether for new fields, setting the oil industry on a par with any other, making it pay no special taxes. But with oil prices soaring and the large fields producing for all they were worth, the potential profits were vast and there was no ideological difficulty in demanding a greater share for the state.

The offshore industries were at their highest level, in terms of both production and their image in the popular culture, during the first seven years of Mrs Thatcher's Conservative government. The giant fields were now on stream, and new fields were constantly being added. Production figures jumped from 80 million tonnes a year in 1979 to 103.4 in 1982, and then up again to over 128 million in 1985 and 1986. (In 1985 and 1986 Britain's consumption of oil was 77.4 million tonnes – leaving a net surplus of 50 million tonnes for export.) Iran's production in 1986 was below 100 million tonnes; Kuwait's just over 60 million tonnes. Britain was producing over 4% of the world's oil. For a very short time, Saudi Arabia, due to deliberate cutbacks designed to maintain the oil price, produced less oil than Britain. Just

Plate 5 Mrs Thatcher inaugurates BP's Magnus oil field, by remote control from Britannic House in 1983 (Photograph courtesy of British Petroleum)

a decade after the first oil price shock, it seemed as if the boot was on the other foot.

Finding rates for new fields in the North Sea were high: the Department of Energy reported that between 1964 and 1984, 942 "wildcat" wells had made 194 significant discoveries. A combination of opportunity and political acceptability was drawing in the oil companies. The majority of effort was concentrated in the central and southern North Sea, but already wells were being drilled in the "frontier" areas: north of 62 degrees north, the northern North Sea, to the west of Shetland, and so on. The Department of Trade and Industry's reports show a continuing level of interest in applications for new licences: for the 1984 9th round, "More applications than ever before were received"; for the 1989 11th round, "This was the highest number of blocks awarded since the 4th round and showed a continuing high level of interest in the UKCS by the oil and gas industry."

Al Alvarez's book *Offshore*,[13] first published in 1986 and a reportage by a poet and journalist, is both a homage to the size and scale of the industry and an attempt to show its human face. It talks in awed tones of the vast scale of the platforms, the money involved, the amounts produced. Alvarez visited the Brent field as a guest of Shell at a time when it was at the forefront of technology and the height of its production. He saw some of the busiest areas of the North Sea and recorded a worker's description of the giant platforms on the major fields as "some of the most expensive pieces of real estate in the world".

Alvarez talked to many offshore workers and managers and records their strange, introverted, isolated life, tempered by their fascination with the technology they were using. The Reverend Andrew Wylie, Chaplain to the Offshore Industry, a post set up in 1986, described another aspect of life offshore: "it is important to dispel the macho myth which surrounds the offshore worker. The ever increasing presence of women in the North Sea will eventually produce a greater awareness of reality; at a light-hearted level their presence and the consequent evidence of aftershave has helped to ameliorate the sometimes excessive masculinity of some installations!"[14]

Both images show large-scale effort taking over from lonely mobile rigs crewed by itinerant Texans; big fixed platforms, highly professional workforces and bureaucratic managers, a sense of technological confidence – and aftershave.

There was one cloud on the horizon: the oil price, having reached a peak of $35 per barrel in 1980, had been declining steadily ever since. In 1986 it almost halved, from $27 to $14 per barrel.[15] The professional managers were now needed to control costs. Most of the extravagance, the high spending and the Wild West atmosphere had to be shaken out. However, it looked as though the industry was maturing. The major investments had been made. The technological problems seemed to have been solved and the heroic days could safely be left behind.

The tragic events of 6 July 1988 were a symbol that nothing could be taken for granted. An explosion and fire onboard the Piper Alpha platform took 167 lives; television pictures of the flaming wreck

Plate 6 "Some of the most expensive pieces of real estate in the world" – platforms in BP's Forties field (Photograph courtesy of British Petroleum)

flashed around the world. Dr Armand Hammer, the chairman of Occidental Petroleum, which operated the platform and owned 36.5% of the licence, flew to Aberdeen for talks with Mrs Thatcher; he donated £1 million to a fund for the victims and their relatives. It was a reminder that the offshore industry could never be taken for granted: it was dangerous. Curiously, the share price of Occidental Petroleum scarcely altered during or after the incident; they were well insured at Lloyds of London, which suffered more than they did. The effect on the industry as a whole was to be considerable.

The Piper Alpha platform was typical in many ways of a large North Sea installation. It had begun production in 1976. In 1987, it produced over 8 million tonnes of oil from the Piper field and exported it via a subsea pipeline to the Flotta terminal. It performed preliminary processing on fluids won from beneath the sea by its own wells and wells in other fields: a mixture of oil, gas and water came up, and the platform separated them. The water was cleaned and dumped back into the sea; the gas and oil were exported down separate lines. The platform extracted condensate (the heaviest, near liquid, components of the gas) from the gas, and reinjected it into the crude oil pipeline for transport to the shore.

The crucial fact is that Piper Alpha was connected to other platforms by one oil pipeline and three gas pipelines. Gas coming from Texaco's Tartan platform – a 12-mile line. Oil coming from Occidental's Claymore – a 22-mile line. Gas going from Piper Alpha to Claymore. Gas exported from Piper Alpha along a 34-mile line to Total's MCP-01 platform and thence to the shore. There was also a 128-mile oil export pipeline to the Flotta terminal.

An initial explosion took place around 10 o'clock that July night on Piper Alpha. Lord Cullen's inquiry into the disaster, published in 1990,[16] concluded that the fire began when night shift personnel attempted to restart a pump which the day shift had shut down for maintenance. Because of poor procedures, the night shift had not been told about the maintenance work, nor that a blank flange assembly, which had been fitted to close the line temporarily, was not leak tight. In any event, highly inflammable condensate leaked out into one of the modules on the platform, and ignited. The fire spread rapidly to another module. The main power supplies and control room were put out of action. The fire extinguishing system failed to work. (It had been reported for maintenance some months before, but nothing had been done.) The emergency response plan for the platform was fatally wrong; following it, the offshore installation manager (OIM) led most of his crew into the accommodation module, to await rescue by helicopter or the nearby *Tharos* support vessel. But the accommodation decks and the helicopter deck were shrouded in smoke; it was impossible to land an aircraft, and the heat was too strong for the *Tharos* to approach. The men inside the accommodation module were trapped.

At 2220, the heat of the fire on Piper Alpha ruptured the riser from the Tartan pipeline and ignited the gas inside it. There was no way of closing off this pipeline: no valve on the sea bed. The valves on Piper Alpha were already unusable. Even if the valves on Tartan had immediately been shut, the whole contents of the 12-mile pipeline were now available to feed the flames; it has been estimated that it took some 55 minutes for the gas in the line to burn. At 2250 the MCP-01 gas line ruptured; at 2318, in what an eyewitness described as "the biggest explosion ... of the night", the Claymore gas riser failed. There was no means of shutting off these lines either. Witnesses spoke of a highly pressurised fire from the surface of the sea, like a Bunsen burner. This caused the final structural collapse of the platform. Lord Cullen's report does not record anyone leaving the platform alive after 2250, when a few crew members who had not followed the OIM were literally blown off the platform by the MCP-01 riser explosion.

The implications of the Piper Alpha disaster for the management of safety have already been discussed. The Cullen report contains 106 paragraphs of recommendations. One short, apparently innocuous, section recommends that offshore operators should "demonstrate ... that adequate provision has been made, including if necessary the use of sub-sea isolation valves, against hazards from risers and pipelines". In other words, there must be a means of turning off the "Bunsen burners". Implementing this recommendation was to cost the industry over £580 million in 1989/90; the overall cost of improved safety related hardware was estimated at over £850 million. Large numbers of subsea lines and risers had to be closed while new valves were installed or modified. Partly as a result, UK oil production fell by nearly 20% in 1989. According to Brian Ward, Production Director of Shell Expro, "Concerns about safety also focused discontent among contractor staff, whose incomes had fallen as North Sea activity declined after the oil price collapse, and the industry suffered a series of damaging strikes."[17]

The new safety requirements included the preparation of Safety Cases, discussed in Chapter 2, and placed a heavy administrative burden on the industry. Many companies welcomed the new regulations and had already moved a long way towards implementing them, but the effect of the legislation was to increase both the paperwork and the uncertainty of old and new developments alike. No one quite knew what Safety Cases should contain, or how stringent the HSE's criteria would be; not even the HSE, which was struggling to recruit enough inspectors to handle its new responsibilities. By April 1994, 215 Safety Cases had been submitted to the HSE; the first was approved in January 1994.

In preparing the Safety Cases, the companies had also been forced to re-examine their work practices. Shell, for instance, found that "we hadn't been doing enough maintenance because we didn't realise how quickly equipment could deteriorate in such harsh conditions . . . Then we began to open such things as electrical junction boxes – there are 15 000 on a big platform – and found we had to replace them all. We ended up with platforms out of service for months."[18]

To compound the industry's problems, the price of oil showed every sign of staying low or going even lower. The "marker" Brent crude price slid abruptly down to $14.38 in 1986, and has varied between $14 and $25 ever since.

The cause of this slide was beyond the influence of the British government. Professor Peter Odell had predicted in 1970 that "there need be no physical shortage of oil in the foreseeable future".[19] Few had believed him in the 1970s, but by the late 1980s there was an oil glut rather than a shortage. The price rise to $30 per barrel had tempted much more investment cash into the market. More money, and improved exploration techniques, meant that oil seemed to be coming to light, offshore and on, in almost every country: from Angola to Vietnam. With the collapse of communist rule in Moscow, the enormous potential of Central Asia, Siberia and the far north was suddenly on offer.

The oil industry had acquired another problem. In March 1989 a supertanker, the *Exxon Valdez*, ran aground and spilled 240 000 barrels of petroleum into the waters of Prince William Sound, Alaska. The incident, and the litigation and bitterness that followed, have become part of popular history; and the oil industry focused its attention on how to safeguard against environmental damage. (It is interesting to compare the *Exxon Valdez* case with the explosion at the Union Carbide plant in Bhopal, India, in 1984. The incident in India killed some 2000 people and seriously injured hundreds of thousands more. Yet the oil spill in Alaska, which caused serious environmental damage but no human injury, led to much higher economic penalties, in fines and litigation, for the company concerned. This may partly be because one incident occurred in the USA and the other in a less litigious country; but the Exxon incident showed clearly the strength of feeling for the environment with which the world was entering the 1990s.)

Britain, too, was facing changes. Mrs Thatcher had gone, overthrown in a party coup in 1990. Her successor, John Major, was believed to have been chosen because he was more in tune with popular feeling and more likely to win the next election (as indeed he did). "Thatcherism" was coming to be synonymous with wealth at all costs, and this was no longer in fashion. The wealth itself no longer looked so secure, for one thing. The legendary British institution, Lloyd's insurance market, had shown signs of profound crisis by 1988, in part due to the Piper Alpha disaster, but also to fraud and the obvious greed of a few people. The world stock market crashes of "Black Monday" in October 1987, though not unique to Britain, had dented confidence badly – the internationalisation of stock and money markets had been encouraged in Britain and London was one of the three world money market centres. The mood of the 1990s was disillusioned, more "caring" perhaps, but also more defensive: times were getting harder and job markets less certain. Britain was also becoming more closely bound up with the European Union which, if nothing else, made the administration of government in London a more complicated and cautious affair.

REFERENCES

1. A Jones, *Oil, the Missed Opportunity*, 1981.
2. F Errol, quoted in C Harvie, *Fool's Gold*, Hamish Hamilton, 1994, p 85.
3. Tony Benn, *Arguments for Democracy*, Penguin, 1982, p 146.
4. *Offshore* magazine, April 1994, p 40.
5. The Scottish angle is covered with depth and commitment in C Harvie, *Fool's Gold*.
6. D Smith, *The Rise and Fall of Monetarism*, Penguin, 1987, p 65.
7. For instance, C Harvie, *Fool's Gold*, Chapter 6; and J D Davis, *High Cost Oil and Gas Resources*, Croom Helm, 1981, Chapter 5.
8. J D Davis, *High Cost Oil and Gas Resources*, p 43.
9. See P R Odell and K E Rosing, *Optimal Development of the North Sea's Oil Fields*, Kogan Page, 1977.
10. Stamping out oil, *Petroleum Review*, December 1992.
11. Department of Trade and Industry, *The Energy Report*, 1995, Volume 1, Foreword.
12. A Jones, *Oil: the Missed Opportunity*, p 161.
13. A Alvarez, *Offshore: A North Sea Journey*, Sceptre, 1987.
14. The soft side, *Petroleum Review*, January 1992.
15. *BP Statistical Review of Energy*, 1994, p 12. Prices are for Brent crude.
16. *The Public Inquiry into the Piper Alpha Disaster*, HMSO, 1990.
17. *Shell UK Review*, March 1995, p 25.
18. B Ward, then Shell UK Expro Director of Production, *Shell UK Review*, March 1995, p 25.
19. P Odell *Oil and World Power*, Penguin (4th edition), 1975, p 226.

4
Now – the mid-1990s

It comes as a surprise, given the gloom in the offshore industry at the beginning of the 1990s, to find a mood of relative optimism, and record production, in the mid-1990s. The oil price has not improved; most UK production still comes from the same "mature" fields; and government policies are still much the same. What has changed is the attitude of the oil companies: they have quietly abandoned the once axiomatic idea that the North Sea had to be a "high-cost" province, and have successfully found lower-cost ways of exploiting it.

The UK's offshore industries are now producing more than ever before. In 1994, the total of oil (126 706 thousand tonnes or an average of some 2.54 million barrels a day) and gas (70 bcm) production was a record. The oil sold for £9.5 billion and the gas for £3.8 billion, and between them they brought in £1.6 billion in taxes for the British government.[1] According to Charles Wardle, then Undersecretary of State for Energy, these figures would make a positive contribution to the UK's balance of payments of some £4 billion – the highest such contribution since 1985. This was despite the fact that the industry now has "one of the lowest oil tax regimes anywhere in the world".[2] (Despite this, UK production was overtaken by Norway in 1991; Norwegian oil production in 1995 was expected to reach 3 million barrels a day, while UK oil production stood at about 2.3 million barrels a day.)

The UK Offshore Operators' Association (UKOOA) claimed that in 1992 the industry invested over £10 billion: £5.3 billion on new developments, £3.3 billion on operating existing developments, and £1.5 billion on exploration.[3] According to the British Department of Trade and Industry, "throughout the 1980s and early 1990s [UK industry] has consistently supplied about 70% of the market", employing some 300 000 people with 27 300 of these working offshore.[4]

Tim Eggar, Minister of State for Energy, boasted that "if you stacked the oil drums [that had been produced in the 20 years since first oil in 1975] they would reach the moon and back a dozen times. And they said it wouldn't last."[5]

The British government could be expected to put a favourable spin on the figures, but it was not alone and this was not just hype. According to many experts, UK waters are a good place to be. The rate of new discoveries continues to be high. In 1994, according to figures prepared by Petroconsultants,[6] more crude oil and condensate were found in new field "wildcat" wells in UK waters than anywhere else in the world. (For the period 1989–93, the UK came second only to Saudi Arabia; for 1984–93, third to Venezuela and Brazil.) In gas discoveries, the UK rated slightly lower: sixth in 1993, fourth in 1989–93, third in 1984–93. Petroconsultants estimate that only one country (Brazil) had more than replaced its production by new discoveries over the 10 years 1984–93. Only three others had replaced more than 75% of their production during this period: they were the UK, Colombia and Angola.

There have been hiccups. The number of wells drilled in 1993 was down, partly because of changes in Petroleum Revenue Tax, which mean that there are fewer opportunities for offsetting exploration expenditure. This change undoubtedly caused some smaller operators to rethink their programmes, and some drilling was cancelled or postponed as a result. The number of exploration and appraisal wells drilled in 1993 was only half that drilled in 1990. Drilling contractors merged or pulled out of UK waters; rigs were "stacked" idle, and day rates plummeted in what is probably the most cyclical part of a cyclical industry.

By 1995, however, there was a shortage of drilling rigs again, and projects were being delayed as a result. Analysts expected the number of wells to pick up again over the next few years, if only because of "obligation drilling" – wells which companies are committed to drill under the terms of existing licences.[7] Capital investment is predicted to remain stable over the rest of the decade, if at a slightly lower figure.

In 1995 there were 73 offshore oil fields and 53 offshore gas fields. Over 60 probable new projects have been identified for the next few years, although many of these are developments of existing fields.[8] In the same year 24 new fields received approval, the highest ever in one year. Expectations have also changed. In the early days, only about one-third of the oil in a reservoir might be recovered; in the mid-1990s at least half can be extracted. The smallest viable field for commercial exploitation was around 100 million barrels; now it is around 10 million barrels. Production has reached a peak of some 3.5 million barrels per day of oil and gas, and is expected to continue at over 2 million barrels per day for 10 or 20 years.[9] The time taken from discovery of a field to production has fallen: from six years to three for a large field, and much less for smaller and easier projects.

It would be easy to dismiss the UK as just a mature province, with little excitement left. To some extent this is true; the traditional North Sea areas have been very extensively explored, and the likelihood of a major field remaining completely undiscovered in these waters is low. But the UK also has its own "frontier" areas. Discoveries by BP of potentially very large fields west of the Shetland Islands could, one analyst has said, herald a new era in the UKCS and create a new investment boom in this sector.[10] The south-west and the Irish Sea are also being actively explored.

In size, the current North Sea fields range downwards from Brent, operated by Shell and owned by Shell and Exxon/Esso, which produced an average of 238 000 barrels per day in 1994 (just under 10% of UKCS oil production). Oil fields producing over 100 000 barrels per day average in 1994 included Forties, Magnus and Miller (operated by BP), Scot (operated by Amerada Hess) and Nelson (operated by Enterprise). At the lower end of the scale are fields such as Staffa (LASMO) which produced an average of 2379 barrels per day in 1994.

The largest gas field is South Morecambe, operated by British Gas, at 864 000 million cubic feet per day (mcfd) in 1994 (about 7.4% of the UKCS total). Other large gas fields include the Shell/Esso Brent field once more (581 000 mcfd), Mobil's Beryl "A" and "B", which between them produced over 411 000 mcfd, Total's Alwyn North (292 700 mcfd), and Amoco's Leman field (253 800 mcfd).

Once produced, oil and gas are brought onshore by several means. The majority goes through a series of pipelines: some to Sullom Voe in the Shetlands, others to St Fergus in Scotland, and Bacton, Theddlethorpe and Immingham in England. Gas fields in particular are linked by major pipeline and control systems such as FLAGS (Far north Liquid and Associated Gas System, which includes the Brent field), SAGE (Scottish Area Gas Evacuation system, which includes Beryl) and LOGGS (Lincolnshire Offshore Gas Gathering System, which includes the Conoco Victor, Viking and "V" fields). Oil pipelines include the Ninian and Forties systems. A smaller number of fields produce without pipelines; in MSR's "Emerald" field, for instance, oil was produced to a floating production facility and transferred by a short subsea pipeline to a permanently moored tanker or floating storage unit, from which it was collected at regular intervals by the purchaser's tanker.

Companies involved in the North Sea fall into several categories. Consortia of companies bid for licences. If these are granted, one specific company must be the legal "operator", responsible for all work on the field. There were some 35 operators in 1995.

A major company such as Shell UK produced some 348 000 barrels of oil per day from UK waters in 1994,[11] nearly 16% of its world total crude production of 2.194 million barrels per day. It is important to distinguish between the amount of oil or gas the operating company produces, and the share of that production to which it is entitled by the consortium agreement. Shell claims to operate about a quarter of British oil and gas production; and indeed in 1994, according to industry statistics, it operated fields producing 525 000 barrels per day of oil and 817 mcfd of gas. Shell UK employs 8000 people (although many of these are engaged in the downstream business, such as refining and marketing petrol); its subsidiary Shell UK Exploration and Production ("Shell UK Expro") operates six gas fields and ten oil fields. Shell UK Expro is itself a subsidiary of the Royal Dutch/Shell group, which has subsidiary companies virtually worldwide. At the time of writing, Shell UK Expro held licence interests in approximately 148 blocks.

The other three of the "big four" in acreage terms are BP, Esso (the UK subsidiary of the US company Exxon) and British Gas. In 1994, Esso produced 347 000 barrels per day of oil and natural gas liquids, and 511 mcfd of gas. This is similar to Shell: the two companies have a 50/50 partnership in most activities. Esso does not currently operate any field in UK waters, although in 1996 it plans to begin operating west of Shetland and in the Irish Sea.

BP produced a worldwide total of 1.262 million barrels per day, of which 429 000 (or 34% of BP's worldwide total) came from UK waters. This was 17% of the UK's oil production. BP-operated fields produced at a rate of 529 667 barrels per day – roughly the same as

Shell-operated fields. BP-operated gas fields produced around 933 mcfd, about as much as British Gas, although of course BP does not have title to all of the gas it produces, while British Gas does.

In other words, BP and Shell are the largest operators, but BP probably has title to more UK oil than any other company. About one-third of Shell's oil production appears to come from one field (Brent). BP has three large fields among its operated portfolio (Forties, Miller and Magnus).

British Gas produced almost all (350 bcf) of its 362.6 bcf of gas from UK waters. Nearly 90% of this came from the Morecambe Bay fields. British Gas is a special case: having developed as a national monopoly, until recently it was not permitted by British law to explore or produce outside UK waters. It is also one of the very few producers to own field licences outright, taking 100% shares without any partners in the Morecambe, North Morecambe and Rough fields. In this sense it operates in a similar way to some national oil companies.

Operators may also be much smaller: for instance, Midland and Scottish Resources plc (MSR) employed 114 people in 1993 and produced 19 300 barrels of oil per day from the Emerald Field, its only producing activity, and the source of almost all its revenue in that year. MSR began production of the Emerald Field in August 1992, but despite additional wells drilled in 1993, the field did not sustain the expected levels and the MSR group went into a voluntary arrangement with its creditors in 1994. The field was produced by one of its subsidiaries until 1996.[12]

Ownership of a licence is frequently complicated, and may change repeatedly. For instance, the Ninian partners as listed in 1994 were Chevron, operator (23.63%), Murphy Petroleum Ltd (13.82%), Neste North Sea Ltd (4.25%), Oryx UK Energy Company (29.54%), Ranger Oil (UK) Ltd (15.81%) and Sun Oil Britain Ltd (12.94%). It is not necessary to be an operating company to own shares in a licence. In judging licence applications, the Department of Trade and Industry looks for technical expertise from the operator and financial weight from the partners. All the Ninian partners are "oil companies", in the sense that all are operators somewhere in the world, if not in UK waters.

Other companies may also have shares in fields. Many of them are the so-called independents, a term which embraces a range of companies. Some are relatively small companies who buy small shares in fields, have little intention of operating, and function as a specialised investment trust. The number of these companies is rapidly diminishing as they merge or are bought out. For instance, in 1991 Union Jack Oil plc, since bought out by Ranger Oil (UK) Ltd, had a 0.75% interest in the Forties field and a 0.6857% interest in Claymore. Union Jack Oil acquired the Claymore share when it bought up another tiny independent, Berkeley Resources Ltd, for £3 million. Union Jack Oil had interests in 19 exploration licences, as a member of various consortia, but its sales in 1990 averaged under 1400 barrels per day – less than 0.5% of Shell's UK production. Union Jack Oil had no full-time staff; it obtained administrative services and technical advice from Ranger, which in 1991 owned 21% of the company. Other shareholders included investment trusts, pension funds and insurance companies.[13]

Other independents may only be shareholders in UK fields, but operate fields elsewhere in the world. An example is Hardy Oil and Gas; according to its 1995 report, this company's total production was 16 134 barrels of oil equivalent per day, roughly 4.5% of Shell's UK production of oil. At the time of its report, Hardy had interests in the UK, Netherlands, Australia, Canada, USA, Pakistan, Libya, Namibia and Algeria. The group employed an average figure of 127 people during 1995. A larger independent is LASMO; worldwide this company had an average daily production of 173 000 barrels of oil equivalent per day, of which 66 500 came from UK waters. (Roughly one-third of LASMO's UK production was gas, and two-thirds oil. This rate of production from UK waters is about 19% of Shell's UK rate.) LASMO employed some 1400 people in 1994, of whom 313 were employed in the UK. It has major interests in Indonesia and the Netherlands where it is also an operator, and in North Africa, Pakistan, Colombia, Vietnam and elsewhere.

There is also active trading in independent companies themselves. Aran, an Irish independent with west of Shetlands acreage, was acquired by Statoil; and Goal Petroleum is the subject of a takeover bid by Talisman at the time of writing. Other attempts have fallen through, such as that by Enterprise to buy LASMO. It must, however, be a nervous period for the management of smaller independents.

Independents are usually not "integrated" companies, that is, they explore for and produce oil and gas, but they do not refine it or market the products. The larger companies, so-called majors such as Shell, BP, and Esso/Exxon, cover the full range of petroleum activities. The economics of the two types of company are fundamentally different, which is perhaps the main reason that both have survived. "Majors" have large cashflow from their downstream marketing operations, and may fund upstream activities largely from their own resources. (Shell has the reputation of being the most conservative in its finance policies; its debt ratio is by far the lowest in the industry.) "Independents", with no such ongoing cashflow, tend to finance on a project basis: they borrow money from banks or elsewhere to develop a particular field, to be repaid from the sales of crude oil or gas from that field. Majors need large volumes of crude and gas to satisfy their enormous and expensive refining and marketing networks; they are typically looking for big fields which will require massive investment. Independents are smaller and can afford to look at niche opportunities such as small or difficult fields. As a stock market vehicle, independents are often seen as a play on the oil price; if the oil price rises, so does the value of their shares. For integrated oil companies, there are balancing factors: the value of their production increases when the oil price rises, but their refining and marketing profits fall since crude costs have risen. Majors are often regarded as "blue chip" shares because of their enormous size and relatively wide spread of activities – although there have been some quite dramatic fluctuations in the share prices of individual majors from time to time.

There is undoubtedly a difference in philosophy between "majors" and "independents". The independents regard the majors as bureaucratic, and claim to be able to operate more cheaply and to respond more quickly; but ultimately the majors have the deeper pockets and produce the majority of oil in UK waters and elsewhere.

Other companies may own shares in the consortia; past examples have included ICI. With the opening up of onshore gas markets, there is commercial logic behind major gas users taking shares in the fields from which they may buy their fuel.

However, non-operating membership of a consortium is not a passive investment. Initial development and subsequent operation of an oil or gas field involves considerable expense and critical technical decisions. Partners meet regularly to consider and vote on such options. These may range from an overall field development plan, to later recommendations by the operating company for further work needed if the field does not behave as expected or the operating plan requires alteration. Later on, as production begins to tail off, it may be necessary to drill additional wells and to consider secondary or tertiary recovery techniques to increase the oil or gas recovery. The costs of each option have to be carefully balanced against the likely returns, and the partners need to agree a common position before seeking

Department of Trade and Industry approval. Once new work is agreed, the "cash calls" to finance it may be heavy, and are made on each partner in the proportion of its share in the field. Failure to meet the calls may lead to penalties. So any partner must keep an active and informed view of the management of the field; it may also need to have access to large sums of money in order to optimise its investment, and it has to balance its cashflow carefully.

For any company, there is a general recognition that all field development is expensive and, despite modern techniques, remains something of a gamble. Almost all operators prefer to spread their risks by taking shares in other fields and inviting others to take shares in their own fields. There is a reasonably active secondary market in these shares of fields or developments; companies buy or sell them to adjust their cashflow projections, to balance their portfolio of risks, or to fine tune their production profiles. For instance, a company with plenty of production at the moment, but few good prospects for five years ahead, may buy a share of such a prospect from a company which is having cashflow problems and would rather have production (and therefore income) now than later.

There is fairly active trading in North Sea blocks. As an example, block 3/10b chosen at random was first licensed in 1971. Since then it has been part of four "farm-ins", one exchange of interests between two companies, three takeovers of interest and one withdrawal. In the 20 or so years since it was first licensed, two wells have been drilled on it: one in 1983/4 found gas, and one in 1991/2 was suspended.

This rosy picture of activity in the North Sea in the mid-1990s should not conceal the fact that the industry is very different from what it was even 10 years ago. It is obsessed by cost and competitiveness. Dr Chris Fay, when Chairman and Chief Executive of Shell UK, struck a surprisingly gloomy note:

> We were distracted by new capital projects and over-complex technology. This is disastrous in maturity when margins are slim and major projects much more difficult to justify. Rather you must concentrate on running your existing assets as reliably and efficiently as possible. This doesn't mean just cutting costs. That is essential but only buys a short breathing space. For longer term success you must increase revenues by earning as much as possible from all your assets.[14]

Dr Fay's choice of words ("disastrous", "breathing space") is typical of a new mood of confession and gloom that swept over the industry in the early 1990s. There had been extensive lay-offs of staff, not just in the UK but internationally. Companies had merged or "downsized", or simply gone out of business. The oil price had stayed between $15 and $20 a barrel and showed little sign of trending upwards. Capital expenditure, on the other hand, which had kept below £2 billion since 1983, rose to £2.6 billion in 1990, and above £3 billion in 1991, 1992 and 1993. Net exports of oil, which had been over 40 million tonnes a year between 1983 and 1987, fell to 4 million tonnes in 1990, and in 1991 and 1992 Britain actually made small net imports of oil. The "North Sea" had lost its romance, and to managers it was increasingly becoming a commodity business: one in which the objective was to produce a basic product as cheaply as possible. The abolition of PRT removed a temptation to drill for tax rebates; the Gulf War had shown that the world oil market did not have to be disrupted by Middle East conflicts.

Energy Minister Charles Wardle, announcing the record production figures for 1994, also pointed out that:

> Future output will depend on ever greater effort to maintain competitiveness. We will continue to encourage the Cost Reduction Initiative in the New Era (CRINE) and . . . press ahead with our initiative to ensure that mature areas remain viable by encouraging improved access to offshore infrastructure.[15]

The companies have now set about trying to cut their costs. The objective of CRINE is to reduce development costs by at least 30%, in order to maximise remaining recoverable reserves, make the UK's offshore industry more competitive internationally and sustain UK employment at high levels. CRINE and its future prospects are covered more fully in Chapter 7.

The government too has acted to simplify the procedures for field development approval. The DTI's *Guidance Notes for Procedures for Regulating Oil and Gas Field Developments*, published in December 1993, are designed to streamline previous procedures and in particular to reduce the "red tape' involved in developing smaller fields. After a licence has been granted, a development programme must be submitted by the operator to the Department to support requests for authorisation to develop and produce the field. It is envisaged that the operator's staff and civil servants should work together to 'assess the correct procedures" and "ensure ultimate commercial recovery" before agreement is given for production. Although government has considerable powers – for instance, to set production levels – this is not normally done.

Now – the mid-1990s

Perhaps the best way to look at the North Sea in the mid-1990s is to examine some typical offshore operations.

THE NINIAN FIELD: THIRD-PARTY USE

In 1994 the Ninian field produced its billionth barrel of oil.[16] However, the field is now well past its prime. An exploration licence for block 3/3 was awarded in the fourth (1971/2) UK licensing round to a group led by Burmah Oil, with partners Chevron, ICI, Murphy Petroleum and Ocean Exploration. The discovery of oil was announced in January 1974; a well had struck the third largest oil discovery in the UK North Sea (after Forties and Brent). The reservoir

Plate 1 A development in 1992/3 using the Ninian system: the subsea manifold of Texaco's Strathspey field before installation (Photograph courtesy of Texaco Ltd)

proved to contain about 3 billion barrels of oil, of which 1.2 billion were assessed as recoverable. In the suddenly inflated oil prices of 1974, the field was worth over $30 billion: a major prize indeed.

The field did not respect the boundaries drawn by the British government on the map; it fell into blocks licensed to the 3/3 consortium and the 3/8 consortium. Interests were allocated between the companies concerned by a unitisation process which allocated 70% of the field to the block 3/3 partners and the rest to the 3/8 partners. The companies involved varied from "majors" such as BP to a small Canadian company, Ranger Oil, hitherto involved in onshore drilling in the US and Canada, which had taken a share in the original exploration and had found itself sitting on a fortune. The Ninian field was to project Ranger Oil, and its charismatic founder Jack Pierce, into the world league.

Plate 2 A subsea module of Strathspey being installed, dwarfed by the Ninian central platform behind (Photograph courtesy of Texaco Ltd)

Plate 3 87 miles of subsea pipeline were laid as part of the Strathspey development (Photograph by courtesy of Texaco Ltd)

Burmah Oil, originally chosen as field operator, had got into serious financial difficulties; in January 1975 Chevron took over as operator, in spite of an initial Department of Energy view that it would not accept an American company as operator. The costs of development were indeed massive; estimates for developing the field were £822 million for three fixed platforms, plus £368.9 million for the pipelines and shore terminal. Production, it was expected, would begin in 1978; so these costs had to be carried for a period of three years before any return could be seen.

The Ninian Central platform, weighing 600 000 tonnes, was towed out to sea in May 1978 and positioned in 460 feet of water, ready for extensive hookup and installation work. This was not without incident; in late 1978, industrial unrest led Chevron to remove the entire construction labour force from the Ninian Central platform in one evening. Two helicopters shuttled 500 people to a nearby flotel, despite attempts to obstruct the helideck and a fight in the cabin of one aircraft.

The first oil was produced on 22 December 1978; peak production of 315 000 barrels per day was reached in 1982. By 1984 the 500 millionth barrel had been produced; but as the field began to decline, it took 10 more years to produce the second half billion barrels (the billionth was produced in October 1994, by which time production was down to 66 000 barrels per day). More than 100 separate wells now reach into the reservoir; some are used for water injection, to maintain reservoir pressure and keep the oil flowing.

As the Ninian reservoir drained, attention shifted to surrounding fields and the use that could be made of Ninian platforms to reach them. Ninian partners identified satellite developments which could contain 100 million barrels of oil. Ironically, Ninian's infrastructure is no stranger to other people's oil: the first oil to flow through its pipeline system came from Unocal's Heather platform in October 1978, two months before Ninian's own crude. By 1994 the Managing Director of Chevron UK, Charles Smith, was pointing out Ninian's "strategic position as the entry point for the Ninian pipeline to Sullom Voe for further satellite and third party business"; satellite and nearby developments already using the pipeline system include Unocal's Heather field (1978), Total's North Alwyn and Dunbar, BP's Magnus, and the "Third Party Project" links to Conoco's Lyell (1993), LASMO's Staffa (1992) and Texaco's Strathspey (1993). The Third Party Project was described by Chevron as "a major and unique North Sea project and not a series of minor platform modifications. It has secured Ninian's future for a considerable time to come." The normal workforce on Ninian Central and Southern platforms has been around 150 per platform. Construction work in 1993 to allow third-party use of the facilities led to over 800 people being employed offshore, with 500 housed on a "flotel".

It is interesting to note the interlocking network of shareholdings in Third Party Project fields. Ninian partners include Chevron (which holds 33% of Lyell); LASMO (60% of Staffa); Oryx (33% of Lyell, 6.5% of Strathspey) and Ranger (40% of Staffa). Holdings change hands fairly often and this list is illustrative rather than definitive; however, it demonstrates the inter-related nature of much offshore activity in the 1990s, and the importance of securing access to infrastructure at commercial prices.

THE BRITANNIA FIELD: A JOINT VENTURE MANAGING AN ALLIANCE

The Britannia field is intended to start production in 1998; at peak it is expected to produce 740 million cubic feet of gas and 70 000 barrels of condensate per day. Reflecting the reduced monopoly of British Gas, Britannia is the first major gas reserve to be sold entirely to the independent gas market. At an estimated 2.6 trillion cubic feet, it is the largest undeveloped gas discovery remaining on the UKCS and one of the more exciting developments in the central North Sea.

The Britannia development is a mixture of traditional and modern approaches. It is centred on a heavy steel platform (weighing some 48 000 tonnes) with accommodation for up to 140 people. It will export condensates by a pipeline joining in to the Forties pipeline system, and gas by pipeline to St Fergus. Rather than build a second platform, the development makes provision for one subsea satellite production manifold; in 1988 it is expected that nine wells will radiate from the platform, and eight from the subsea manifold, which will be controlled from the platform.

The development is made more complex by its commercial organisation.[17] There are seven partners, of which the two largest are Conoco (42.41%) and Chevron (30.20%). These two have formed a further company, Britannia Operator Ltd, which they own and staff jointly, to undertake the legal and practical roles of field operator. The development of the platform "topsides" – that is, the design and construction of the platform, installation and commissioning – is conducted by a group of seven companies called the "Britannia Topsides Alliance". These include Britannia Operator Ltd (BOL), AMEC as designer and for the hook-up and commissioning stages, Trafalgar House as Deck Fabricator, SLP Engineering Ltd (Lowestoft Yard) for accommodation design, procurement and fabrication, and others. The motivation behind the alliance is to reduce capital expenditure ("capex") and operating expenditure ("opex") by removing duplication and inefficiencies and taking advantage of a new contract structure (see Chapter 7). According to BOL, this "aims to remove the barriers which are sometimes built between client and contractor, and ensure alignment of objectives and processes". It includes "aligned goals and objectives" and "non-adversarial contracts".

The reasoning behind this alliance is to some extent the need to remove the costs of duplication. Rather than having a project team in Chevron and Conoco, and in each major contractor, there will be one senior management team, which will report to an alliance board.

The alliance is also a means of sharing the risks and the rewards. The contracts between BOL and the alliance members are highly complicated. They include a payment structure which involves a mixture of agreed lump sums, reimbursement of direct costs, and incentives – the latter based partly on the individual contractor's own success in meeting cost targets, and partly on the success of the project as a whole. (This method of alliancing may not, one suspects, enable companies to downsize their accountancy departments; working out the payments seems unlikely to be a simple matter.) There is an "alliance charter" which is not legally binding, though members undertake to operate within its terms. It covers the alliance's objectives and principles, the behaviour and rights of members, the operation of the alliance, and administrative principles. BOL talks of a "spirit of openness and trust" in the alliance. Again, developments are becoming more inter-related, in this case by closer connections between the companies involved.

THE ANGLIA FIELD: A NICHE EXPLOITED, NETWORKS OF OWNERSHIP AND INFRASTRUCTURE ACCESS

The Anglia gas field was first discovered in 1972. It is some 30 minutes flying time from Great Yarmouth, in shallow water. Due to low gas prices and lack of markets at the time, it was not until late 1990 that Ranger Oil (UK) Ltd drilled the first appraisal well of this field. When they decided to develop it, they achieved some impressive technical records: they broke, and then rebroke, the UK record length for a horizontally drilled well section. A small platform was installed over the well-heads and full production began in November 1991, only some 13 months after UK government approval was given for the development. Ranger owns just over a third of the operation; its partners are Conoco and Amerada Hess.

The Anglia Alpha platform is a small "not normally manned" installation. It weighs only some 2000 tonnes, and is almost dwarfed by the helicopter landing deck on top of the structure. There is rudimentary overnight shelter for use in emergencies only; but the platform is remotely controlled and operated, and is visited only once every two or three weeks for maintenance or development. The Safety Case agreed with the HSE does not permit visitors to stay overnight, except in emergency; like the very early offshore drillers working from piled structures within sight of the beach, any operatives on Anglia travel out in the morning and come back before nightfall, taking packed lunches with them.

Small though it is, the Anglia platform is part of LOGGS (Lincolnshire Offshore Gas Gathering System), run by Conoco, a network of platforms and pipelines which collects gas from several North Sea blocks, gives it limited processing offshore, and transports it under the sea back to Conoco's onshore terminal at Theddlethorpe. Day-to-day control of the platform, and of the amounts it produces, is through LOGGS; all its activities are remotely run from a nearby platform or from the LOGGS control room onshore. In other words, the field is run day to day by a company (Conoco) which operates its infrastructure, acting as agent for the "operator" (Ranger) and through Ranger for the partnership (including Conoco itself) which holds the field licence; its gas is brought onshore and treated along with gas from several other fields.

Anglia has also benefited from changes in the way gas is sold in the UK. It is one of the first North Sea gas fields from which production is sold directly to distributors and end-users, rather than via British Gas. Anglia gas was initially contracted to two onshore gas marketing companies, and then to National Power plc for electricity generation on a long-term contract.

Ranger is a relatively small company and Anglia represents a significant part of its hydrocarbon production. At 22.5 million cubic feet per day average in 1995, it amounted to 16% of the company's worldwide gas production. The gas produced from the Anglia field was sold at an average price of $3.18 per thousand cubic feet – i.e. gross revenue of about $25 million for the year.

To put Anglia's production in perspective, in the same year Ranger earned $20 million from transportation and processing facilities provided by the Ninian field, in which it has a network of shareholdings. (According to its 1990 report, these were 11.5% of the Ninian field, since raised to 15.8%, 7.1% of the Ninian pipeline, and 1.9% of the Sullom Voe terminal). That is, Ranger earned roughly as much from its share of the charges made by the Ninian partners for handling other people's oil as it did from selling its own gas from Anglia.

These figures demonstrate again how the offshore industry is changing. Small fields can be operated on a day-to-day basis by companies which are not the legal "operators", but control them as part of a larger integrated system. The revenue from ownership of control, throughput and processing facilities can now be equal to that from a small field.

THE FRIGG GAS TRANSPORTATION SYSTEM

At any time, up to one-third of the natural gas used in Britain may come ashore through the St Fergus terminal in Scotland. This is the end of the Frigg Gas Transportation System, operated by the French company Total Oil Marine. The pipeline, at the time the longest of its kind in the world, was laid in the 1970s to transport gas from the Frigg fields, in the Norwegian sector of the North Sea, to the UK. (Because of a deep trench to the east of the Frigg fields, it is considerably easier to run a pipeline west from Frigg to the UK, even though the field is in Norwegian waters.) There are two parallel lines, each 32 inches in diameter and running 225 miles. Throughout the 1980s and 1990s the system was extended to include other fields: Tartan (1981); North East Frigg (1983); Odin (1984); Petronella and Alwyn North (1987); East Frigg (1988); Rob Roy/Ivanhoe (1989); Bruce (1993). As the system has grown, its components have changed. The enormous concrete platform known as MCP-01 was originally built, half-way along the pipeline between the Frigg field and the coast of Scotland, as a gas compression platform. In 1992 it was converted to "not normally manned" operation and is now used as an entry point for gas from the newer system participants. It is operated by remote control from the St Fergus terminal.[18]

The system is made more complex because the Frigg line only handles gas; oil produced, say, by Total's Alwyn North platform is transported via the Ninian pipeline system to Sullom Voe. Some of the fields served by Frigg, such as Tartan, produce mostly oil; without the Frigg system, it would be uneconomic for them to produce gas as well. A field such as Alwyn North is therefore producing oil and gas, using both pipeline systems. This links it to other companies operationally and commercially. Operationally, it is dependent on the pipeline system operators to control overall flows in the system; commercially, its production is commingled with the products of several other fields and it is relying on the pipeline system operator to treat its products and to provide an accounting and control system which allows it to sell an equivalent amount of natural gas liquid to that which went in to the line. (The actual production from each field is not kept separate from the others.) Its profitability is also partly dependent on the charges made by the system operators for throughput and processing services.

Once the gas reaches the St Fergus terminal it becomes part of another web of pipelines. Some fields in the system produce gas which is "richer" or "heavier" than the original Frigg gas, which is largely methane, as used for domestic and industrial heating. As well as methane, the richer gas contains a higher proportion of heavier hydrocarbons: natural gas liquids or NGLs such as butane. (These "liquids" are liquid only below minus 45 degrees Centigrade.) NGLs are separated out at St Fergus but not processed there. While the methane is sold directly into the British Gas pipeline system for transportation around the UK, the NGLs are sent by pipelines to BP at Cruden Bay or to a nearby Shell terminal which are more adapted to process them.

Thus, NGLs from Tartan (say), which is operated by Texaco, may come through the Frigg system operated by Total, then pass to Shell or BP for treatment before being used. Production rates will depend on the operation of the systems and the other system users. The development profile of a field (for instance, whether it is worth developing an associated gas reservoir) may depend on the availability of a suitable system.

ABANDONMENT

The problem of what to do with ageing North Sea structures jumped to the forefront in mid-1995, when Shell UK's attempts to dump its Brent Spar facility in the Atlantic were sabotaged by protesters from Greenpeace and other organisations. Early in 1995 Shell removed the Brent Spar, a storage and loading facility which had become redundant, from its place in the Brent field. The structure is a large one, weighing 14 500 tonnes, 141 metres high and with a draught of 109 metres. All sides admit that the Spar contains some residues in the form of oily sludge, which may contain heavy metals and radioactive wastes. Opinions differ on the quantities involved and their impact.

As required by the British government, Shell undertook environmental studies and prepared a detailed case for the disposal operation. Shell argued that the best option was to sink the Spar in the Atlantic, 240 kilometres out from the west coast of Scotland, in some 2000 metres of water. This argument was accepted by the British government in February 1995, and approval given for the operation to take place. While it was still very much feeling its way at the time, the government had already approved eight other abandonment programmes – although all of these had involved total removal of the facilities from the sea bed.[19]

The international legal framework for abandonment is set by guidelines of the International Maritime Organisation (IMO), which it is UK government policy to observe. This requires that all installations should be entirely removed, once they are abandoned, except where a case-by-case assessment shows that it would be too expensive, difficult or dangerous to do so. No exceptions are allowed for platforms in less than 75 metres of water and weighing less than 4000 tonnes, which must still be removed entirely. (After January 1998, this limit is extended to less than 100 metres of water.) Installations in depths greater than 100 metres, or weighing above 4000 tonnes, may be partially removed providing that there is at least 55 metres of clear water above any submerged remains, in order to reduce any hazards to navigation. (After 1 January 1998, the design of all installations must enable complete removal to be a feasible option.)

Dumping of wastes or other materials at sea is controlled by the London and Oslo Conventions of 1972 and the Paris Convention of 1992. These proscribe the dumping of specific substances such as mercury and cadmium, and restrict such things as marine incineration and the dumping of sewage sludge. Some of the proscribed substances may be found in oil sludge, or used on offshore installations, although this had been addressed in Shell's disposal case. (The British government has resisted or delayed adoption of some measures under the Oslo and Paris Conventions, such as the dumping of sewage sludge and some industrial wastes. It will be recalled from Chapter 2 that Greenpeace in particular has been very active in opposing such dumping. The battle lines have therefore been clear for some time.)

There was opposition from two sources to the decision to sink the Spar in the Atlantic. First, some of the governments participating with Britain in the Oslo and Paris Conventions argued strongly against the British government's decision to approve Shell's proposal. Secondly, environmental groups, spearheaded by Greenpeace, mounted a skilful publicity campaign from the end of April 1995, landing activists by helicopter to occupy the Spar. Public protests were orchestrated in many European countries, including boycotts of Shell filling stations and products. Some individuals took these to criminal lengths: a Shell filling station in Hamburg was fire-bombed. The environmental groups argued that the disposal of the Spar would release large amounts of toxic material which would seriously damage the Atlantic environment. Shell replied that any damage would be limited to a small area of the sea bed, and that the risk to health and the environment was less than if the facility and its contents were disposed of on land. (Ironically, once it had succeeded in its objectives, Greenpeace apologised to Shell UK and admitted that some of the scientific arguments it had used in its campaign had been incorrect.)

Against this background, and while the British government held firm and continued to support the Atlantic disposal option, Shell abruptly announced in June 1995 that it had decided to abandon its plans for deepwater disposal of the facility, and would seek a licence for onshore disposal following a further review of the options and the preparation of a suitable alternative disposal plan. It is not entirely clear why Shell changed its mind; but the cost of lost sales of gasoline and other petroleum products from some of its continental outlets was widely believed to have tipped the balance. Annoyed at the *volte face*, the British government could only comment that Shell would have to work "extremely hard" to get a licence for onshore disposal.

This has left a degree of uncertainty among operators, especially those with platforms nearing the end of their lives, to which we shall return in Chapter 9.

SAFETY

Offshore safety remains a major issue. The first Safety Case to be accepted by HSE, in February 1994, was for the Hamilton Oil installations in the North Ravenspurn gas field in the southern North Sea.[20]

(Hamilton has since been renamed after its new Australian owner, BHP.) Preparing the Safety Case reportedly cost about £1 million, to cover a processing and accommodation platform and three satellite well-head platforms. The HSE Offshore Division, with about 350 officers concentrated in Aberdeen, Bootle and Norwich, faced the enormous task of studying and approving the Safety Cases for all UK fixed installations and any mobile drilling units working in UK waters, before its deadline of November 1995. (While, of course, also coping with a continuing stream of new Safety Case submissions for new fields and projects.)

Describing the Safety Case philosophy in *The Chevron Magazine* in early 1995, the Chief Executive of the Offshore Safety Division, Roderick Allison, said:

> The Safety Case now becomes the basic building block in constructing a new set of modern regulations. Operators have to identify the important hazards and write down how to deal with them. Once that has been done it becomes possible to adopt a much more open approach in which we can go for goal-setting regulations which say "this is the objective – now you set out the means for delivering it . . . The buck always stops with the operator – he's the one who creates the hazards, and he's the one who has to deal with them . . . our role has become to check that the operator is fulfilling the requirements of his own safety case as well as those of the law".[21]

New legislation is being introduced, based on the "goal-setting" philosophy but going beyond it to some extent. It is partly driven by the need to implement the European Union Council Directive (Extractive Industries) (Boreholes) of 1992 (see Chapter 9). Two pieces of legislation have so far been introduced: they are the Offshore Installation and Pipeline Works (Management and Administration Regulations) 1995 or MAR, and the Offshore Installations (Prevention of Fire and Explosion, and Emergency Response) Regulations 1995 (known to the industry as PFEER, pronounced "fear"). A third, the Design and Construction Regulations, was still to appear at the time of writing.

The new legislation is written in layers, with different degrees of detail and of force. For instance, PFEER (the Statutory Instrument) is a short piece of legislation. It talks in general terms. However, it is published with a second layer, an Approved Code of Practice or ACOP, which expands the bald statements with suggested ways of implementing them. The ACOP does not have the same force of law as the regulations themselves. While an operator must comply with the provisions of PFEER, it is free not to follow the ACOP recommendations if it can demonstrate that it has a better way of achieving the objectives of the PFEER legislation. In addition to the ACOP, the PFEER regulations are published with a third layer of "accompanying guidance", which has still less legal force than the ACOP, but which has to be taken seriously by the operator and thus acquires a quasi-legal status.

The PFEER regulations themselves are very broad. Thus, for instance, Regulation Eight, "Emergency response plan", says that the Duty Holder must prepare and keep up to date an Emergency Response plan, covering arrangements and procedures for an emergency. In preparing it, he should consult "persons who are likely to become involved in an emergency response". It stipulates that his plan must be available to all on the installation, who must be given "such notification of its contents as are sufficient for them". The plan must be tested as often as appropriate. Everyone on the installation must conform to the procedures in the plan, as far as is practicable. These (paraphrased and summarised) are the strict requirements of the law.

The ACOP expands on this. It specifies certain bodies who should be consulted in the preparation of the plan (such as HM Coastguard, and operators of related pipeline or installation owners). It defines in a little more detail what the plan should cover (such as who does what, both onshore and offshore, from when the emergency is first detected until it is over). It adds a requirement that the plan should be concise and readable. It expands slightly on the frequency with which the plan should be tested. It adds a requirement to review and rewrite the plan in the light of experience gained in tests or real incidents.

The third element is the accompanying guidance. This adds still more to the list of organisations who should be consulted in drawing up the plan (including the police, and possibly local authority fire services). It expands the definition of "arrangements and procedures", for instance to say that these should include the chain of command, arrangements for communications, and actions to be taken in response to specific emergency scenarios (although it does not specify what these scenarios should be).

Some see this as a creeping return to prescriptive legislation. The Safety Case legislation allowed the operator to set its own rules, provided always that they were accepted by the HSE. The new regulation/ACOP/guidance pattern begins to become more specific about what should be done. There has been considerable discussion between the government and UKOOA about the contents of the ACOPs and guidance notes.

The move to "non-prescriptive" legislation was a realistic recognition that technology and operating methods change. Under earlier safety legislation, platforms might be inspected to determine the number of (say) lifebelts. A given number of lifebelts was prescribed by law; and even where lifebelts were no longer seen as the best safety measure, the operator had to have them. More dangerously, if the operator had sufficient lifebelts in good order, then it was under no obligation to have an alternative system which might be widely accepted to be far better. Safety could easily become a matter of ticking off items on a list, requiring no thought from the operator. Taking the example of the Emergency Response plan, earlier HSE Guidance (the *Offshore Emergencies Handbook*) contained a list of specific emergency scenarios which the Emergency Response Plan should cover. This is no longer current. The operator has already defined a list of potential serious risks in its Safety Case, and it is logical to expect that its Emergency Response Plan should be based around this list.

The new system attempts to put the responsibility for identifying the most appropriate precautions firmly on the operator's shoulders. But clearly, any system in which the operator is responsible for setting its own safety standards is potentially open to abuse. The balancing role taken by the HSE is first to review the Safety Case containing these precautions and safety standards, to ensure that they are reasonable, but not to enforce any single means of reducing risks. In later legislation such as PFEER, the HSE has moved closer to suggesting specific measures.

However, the regulation does suggest some fairly specific measures and standards. For instance, the regulation calls for "adequate instruction and training in the appropriate action to take in an emergency". The ACOP expands this to include training in personal survival, installation-specific induction training, and "training based on the emergency response plan". In practice, this requirement is further codified by the UK Offshore Operators' Association (UKOOA). UKOOA works through a system of committees made up of experts from the operating companies. On the advice of its Safety Committee, UKOOA produces a list of recommended levels of training, say in "personal survival" and the individuals to whom these levels are appropriate. So, for instance, a one-day offshore survival and fire-fighting course, held at a recognised institution and repeated every four years, is suggested for anyone who spends a night offshore. Most operators adhere to these recommendations and will not allow anyone offshore without the appropriate certificates.

The law is therefore a balance between self-regulation by the operator, industry self-regulation by UKOOA and other bodies, and the layers of guidance, ACOP and regulation issued by the government. It is also a balance between British practice of consultation, and the UK's obligations to implement European directives, which are not of course geared specifically to UK offshore conditions. There is likely to be a continuing creative tension between the different approaches – between advocates of prescriptive legislation on the one hand and of controlled self-regulation on the other. While this does not lead to simple, clear-cut positions, or minimise the amount of paper used in regulations, guidances, codes, recommendations, and so on, it is probably the most effective means of controlling such a technically complex and varied sphere of activity.

FUTURE DEVELOPMENTS

In 1992, according to Dr Chris Fay, Managing Director of Shell UK, over 40% of UK oil production and one-third of UK gas production came from fields that had come on stream during the 1970s.[22] These fields are becoming more expensive to operate as platforms and equipment age, reservoirs decline, and the proportion of water produced (instead of oil or gas) grows. The next chapters describe some ways in which the North Sea is changing, and offer alternative views of what its future may be.

REFERENCES

1. Unless otherwise stated, statistics in this chapter are from *The Energy Report*, April 1995, published by Department of Trade and Industry/HM Stationery Office.
2. Tim Eggar, when Undersecretary of State for Energy, *Petroleum Review*, March 1994, p 109.

3. Quoted in *Petroleum Review*, April 1994, p 174.
4. *The Energy Report*, April 1995.
5. DTI press notice, ref p/95/392, 14 June 1995, p. 30.
6. Quoted in *Oil and Gas Journal*, 6 June 1994.
7. Norland Consultants, quoted in *Euroil*, April 1994, p 14.
8. Ibid.
9. See talk by John Browne of BP to Royal Institute of International Affairs, 2 November 1994. See also Chapter 8.
10. Wood Mackenzie, quoted in *Euroil*, April 1994, p 14.
11. *Shell UK Review*, March 1995, p 10.
12. MSR plc annual report for 1993, issued 6 July 1994.
13. Union Jack Oil plc, annual report for 1990.
14. *Shell UK Review*, March 1995, pp 4–9.
15. Quoted in *Oil and Gas Journal*, 20 February 1995, p 100.
16. "The Ninian Billion", special issue of *The Chevron Magazine*, 1994.
17. Information from two Fact Sheets produced by Britannia Operator Ltd, *Britannia: Britannia Operator Ltd* and *Britannia: the Topsides Alliance*, both undated.
18. Information from brochure published by Total Oil Marine plc in 1992, *Frigg Gas Transportation System*.
19. Tim Eggar, Minister of State for Energy, quoted in *Petroleum Review*, March 1995, p 116.
20. According to *Oil and Gas Journal*, 18 April 1994, p 31.
21. *Chevron Magazine*, Spring 1995, pp 8–9.
22. From an address to the Aberdeen Chamber of Commerce in October 1992, published by Shell Publicity Services, UKAA/2, Shell Mex House, Strand, London.

5
The Scenario Planning Process

As the previous chapters have argued, the "North Sea" oil and gas industry has changed dramatically over the last 25 years. These changes are continuing, if anything intensifying. However, there is every sign that the industry's success or failure will continue for many years to be of fundamental importance to the British economy. That success or failure will be determined to a large extent by the ability of all concerned to take a clear view of the future.

It was against this background that the Institute of Petroleum (IP) decided to hold a Scenario Planning Workshop to look at the future of the offshore oil and gas industries. We felt that it would be helpful to all concerned, not just to IP members, if we offered some structure for thinking about the future, which they might use in forming or testing their own plans.

This chapter explains what the scenario planning technique is and why we chose it. It also describes what we did, and why we felt that our group was a novel approach to the technique.

It might be useful to begin with a short description of what scenario planning is not. It is not, for instance, an attempt to make predictions. "Fortune telling" of one sort or another is a highly organised activity. It survives not only in the daily horoscope columns of newspapers, but also in far more respectable guises. It usually claims some form of greater insight into the matter in hand: whether this is through an understanding of the patterns of the stars, a "sixth sense", or simply a greater knowledge of the subject or an "inside tip". This feeling may or may not be right, and the "credibility" of the person making the prediction is not always a valid indicator. (As Chapter 1 recorded, Sir Eric Drake, the then Chairman of BP, predicted that there would be no major oil finds in the North Sea only a few months before his company found the Forties field. Presumably he had access to at least as much insider knowledge as anyone else.)

A much more reputable form of predicting the future makes deductions by extrapolation, that is, by assuming that current trends and influences will continue, and then working out what these will bring in the future. If daffodils flowered last spring, then they will do so next year. Most human activity would be impossible without such extrapolation.

Econometric predictions build on this approach. They employ very sophisticated mathematical techniques to analyse relationships between economic factors, and to produce equations which link a range of factors, in order to deduce future behaviour from the past and present. They are especially useful in the shorter term, in cases where there are long leads and lags, such as the oil and gas industries. Econometric techniques can be relatively successful in looking at supply and demand. The upper limits of oil or gas supply can, in theory, be estimated, since new production cannot be turned on like a tap; there are lead times involved in building it up. If you know how many gas projects are under construction today, you have at least a top limit of the number which will be on stream next year. Projects may, of course, be cancelled or suffer unexpected disasters; but if it takes more than a year to build them up, then one that does not exist now will not be on stream next year. Demand can also be estimated, if it is assumed to have a predictable relationship to economic growth, which can itself be predicted using econometric techniques.

However, extrapolation cannot anticipate a major change in the rules of the system: new technical developments drastically cutting down the construction time, for instance. It cannot anticipate new constraints, for instance political upheavals in major gas-producing countries which might nullify much investment and slow or stop projects altogether.

Even if one could find a suitable technique, looking into the future is a dangerous activity. As Cassandra found, it can damage your credibility, even if you are right. To demonstrate this, we began our workshop (held in May 1995 to look at the next 15 years) by looking back 15 years. In order to demonstrate the risks of "fortune telling" and to make clear in our minds what we were trying to do, we asked the question: "If we had held a similar workshop in 1980, to predict what the offshore industry would look like in 1995, what might we have anticipated?"

The history of those 15 years, 1980 to 1995, contained three sorts of factors:

- Factors that any observer in 1980 knew were likely to change, but where the direction and extent of change were, with hindsight, unpredictable. These included the oil price. In 1980 the cost of a barrel of oil seemed set to reach the stratosphere; in 1995 it had fallen to a relatively low historical level and showed every probability of staying there. A "rational" oil price prediction made in 1980 would almost certainly have been three or four times too high. (Professor Peter Odell, who got the pattern if not the actual figures right, was widely dismissed at the time as a "contrarian".)
- Events that an intelligent observer might have predicted in general terms, but not in any specific way. For instance, given the dangers of the offshore world, we might have guessed that there would be one or more serious emergencies; but we would not have predicted the "Piper Alpha" disaster specifically, nor its ferocity and tragic size.
- "One-off" events that we would be unlikely even to have guessed. (These included, for instance, the Gulf War, the destruction of much of Kuwait's production equipment and the enormous pollution which followed – and the fact that none of this had any long-term impact on the oil price.)

Furthermore, we realised that even if we had, through some stroke of luck or some unique depth of perception, predicted some of these events rightly, our predictions would have been useless in 1980. For a start, it is likely that no one would have believed us. Even if some senior industry executives had respected our expertise or our prescience, they could not afford the risk of designing their strategies and betting their companies' future on such specific "off-the-wall" statements. It would have seemed incredible that Iraq could invade Kuwait and be driven out again, with no long-term effect on oil prices. Any attempt at "fortune telling" has a "take it or leave it" implication. In effect, the fortune teller is saying, "this is what I think will happen; if you don't agree, then my forecast is of no use to you".

The advantage of scenario planning is that it enables organisations to look at the future without falling into the trap of fortune telling. Our workshop did not have to make specific predictions, and our audience did not have to take them or leave them. Scenario planning is more concerned with the range of choices and possibilities than with selecting any particular path through them. The objective is to think about the future rather than to describe one predicted future. We were trying to draw a map of the different roads the future may follow in the next 15 years, rather than to give one address at which it will be living in 2010.

Our role was to look at ranges of possibilities: to look at pathways across our map of the future, and then to assess three or four of the most likely even though they led us to different corners of the map. It is almost irrelevant whether in 2010 our ideas will be proved "right" or "wrong". (We covered our bets by producing four different scenarios, so there is a fairly good chance that at least one will bear some relation to reality, and at least three of them will be wrong.)

What matters is that our work should assist the thinking, in the mid-1990s, of those who have to design strategies and shape the future of companies and organisations. If we offer an easy answer ("the oil price will be this; the available technology will be that") it is of limited use. If we offer a range of choices, and a detailed examination of the parameters and drivers behind those choices, the response is, we hope, to stimulate thought. What would we do if this "came true"?

Scenario planning techniques have been in use for a long time. In a modified form, they are part of the basic equipment of every serious chess player. The first 10 (or more) moves of the game of chess have been exhaustively analysed, and a series of preferred pathways selected. These are the "openings" listed in chess books under such romantic titles as "King's Indian", "Evans Gambit", etc. Each is, in effect, a scenario. If I start with this move, my opponent can reply with any one of (say) 20 counter moves. If he (or she) is a reasonable player, he is most likely to choose one of a handful of moves from that 20. As soon as he does, then I can form some idea of which strategy he is taking; and I can select my own set of moves to reply.

Serious chess players dedicate much study to the different "openings". As each one is played over and over again, new "lines" are discovered and analysed. New ways out of difficult situations, new ways of exploiting advantages, are identified, recorded and committed to memory. There are several advantages of this level of preparation. First, it reduces the possibility that the players, under the pressure of time or the strain of competition, will simply miss an obvious line of play or a clear threat. They do not have to work out every move from first principles. Secondly, it allows them to see beyond the tactics of play and to think strategically. Handbooks of "openings" frequently qualify their blow-by-blow descriptions of the sequence of moves with strategic assessments. This opening is safe and conservative; it may not lead to rapid victory, but it leaves no hostages to fortune. That one develops a strong king-side attack quickly, but the middle game requires difficult and complex play. With these assessments in mind, and balancing them against an assessment of your own skill relative to your opponent, you are much better able to look ahead and to choose your strategy.

This is neither prediction nor "fortune telling". A fortune teller would say: "You will win by a checkmate with your queen on your 25th move." A strategist is saying something like: "This game may depend on control of the centre. Any attack on me is likely to be on the king's side, so I must watch that. I am likely to have opportunities to develop my bishops, so I must seek out lines to use them."

A third advantage of studying and learning these chess openings is that the player is more quickly able to spot what the opponent is doing. If my opponent replies to my cautious initial moves with a strategy that is known to be aggressive but high risk, I have a good idea of how the rest of the game will go. A good player should be able to identify the opponent's opening strategy (and indeed name it) within the first few moves; from then on he or she has an idea of how the game will develop, what the opponent's strengths and weaknesses will be, and how to defend against the first and capitalise on the second.

However, chess is not like real life. The possible number of moves available during the opening of a chess game is vast, but it is finite. Players all around the world sit down over and over again, with the same rules, the same forces at their disposal, the same precise, abstract language to describe the moves. Every player has the same objective and the same means to achieve it: to win by capturing the opponent's king.

On the other had, the oil industry, like anything else in real life, is unique. If there are rules, we only know some of them. Neither are there any real parallels in time or space. To some extent we can look at what happened to other industries and other players: are there similarities, for instance, between the North Sea boom years of the 1980s and the early deep mining stage of the South African diamond industry of the late nineteenth century? To some extent we can look at the offshore oil industries of other countries, but even here the "rules" are always different. The reserves are larger or smaller, the governments more or less welcoming, the technical problems greater or less significant, and so on.

Usually our knowledge is historical: we know what has happened over the last few years, but we also know that every so often there is likely to be a "paradigm shift" during which the rules will change completely and unpredictably.

Short-term predictions may turn out to be accurate. Over any longer period, forecasts become increasingly vulnerable to paradigm shifts. A convenient example, once again, is the 1973 rise in the oil price. In *The Death of Economics*, Paul Ormerod argues that:

> We need to abandon the economist's notion of the economy as a machine, with its attendant concept of equilibrium. A more helpful way of thinking about the economy is to imagine it as a living organism. The economy has the ability, when prodded or stimulated, not simply to wobble around a fixed position, but to jump to a different position altogether.[1]

Looking for the sort of events which cause these jumps, Professor Susan Strange describes:

> key decisions which have altered the course of world economic history in recent times, and which have shaped the development of the world economy and determined shifts in the costs and benefits, the profits and losses, the risks and opportunities, amongst nations, classes and other social groups.[2]

(Professor Strange's concept of a "decision" is wider than formal specific decisions: it may also include collective decisions or non-decisions.) These "decisions", or "jumps to a different position altogether" make prediction by more or less sophisticated extrapolation techniques a risky business. If we assume the rules will stay the same for a short period, extrapolation offers answers which may be more or less accurate for that limited time. However, extrapolation cannot detect the "paradigm shift" which rewrites the rules. Chess has the advantage that there are no paradigm shifts; the knights will not suddenly acquire the powers of bishops; the board will not grow an additional row of squares.

In real life, a cynic might say, it is precisely because of the paradigm shifts that the big money is made, or of course lost. Examples abound in industry: the fortunes of Microsoft or Apple would have been very different if the computer industry had remained in the mainframe-dominated world of the 1960s.

A further difference between real life and chess is that there is no single agreed objective which all players in the oil industry are pursuing, nor is the game finite. Governments may want to maximise their income, or to develop their resources cautiously with a view to the longer term. Shareholders may want to make a quick killing; company officials may want to ensure that their own jobs are secure. Environmentalists want to impose tighter controls on activities; trade unions to ensure that their members have good, safe, secure and well-paid jobs; consumers that the price of petrol does not go up; and so on. Whether by the invisible hand of Adam Smith, or the highly visible finger of government, these aims achieve some sort of balance at any given moment – but the balance is always changing.

Even though the abstract techniques of chess cannot be applied to real life, it was recognised that there are advantages to thinking scenarios out in advance. Attempts were made to develop techniques to do so. The Royal Dutch/Shell oil company is one of the pioneers. In the early 1970s, Pierre Wack, of the company's new Group Planning Department, used scenario techniques to examine the possibilities of a change in the oil price. It was becoming increasingly likely that there would be a "paradigm shift": the relatively stable relationships between oil companies and oil-rich countries would break down and the oil price would rise. But when, and by how much, and what effect would this have? Wack spoke of "reperceiving" rather than "forecasting"; his main concern was the intensely practical one of getting Shell managers to think through the possible implications of an unpredictable future, and apply them to their own task-specific planning. Whether as a result of this or not, when the price rises came Shell prospered while other oil companies declined. Shell went on to refine the techniques and to use them more widely.

Another much-publicised use of the scenario planning techniques was in South Africa, by the Anglo-American Corporation. This was described in *The World and South Africa in the 1990s*, by Clem Sunter, a book published in 1987, a time of great uncertainty in South Africa.[3] Nelson Mandela was still in prison and a white National Party government in power. The rules of apartheid were still rigidly enforced. But it was obvious that change had to come; the old system was becoming less and less viable, and even the most traditionally minded whites were willing to admit, privately, that one day there would be a black State President. Anglo-American were first interested in the technique of scenario planning by a presentation made to them in 1982 by Shell. In 1985, using a "hand-picked team of world experts", they set about producing scenarios for the future of South Africa.

This was a very appropriate forum for the technique. Many South Africans were focused on the "either/or" dichotomy: either there will be a black government which may lead to a breakdown of white society, or there will be a right-wing backlash and the whites will try to hang on to power. Sunter's book allowed them to consider two scenarios: the "High Road", in which South Africa changes, becomes more acceptable to the outside world, and sanctions fall away; and the "Low Road", in which the reverse happens. Sunter's book explores the likely ways in which each could develop, and the probable results, both political and economic. One of his most powerful arguments was that there could be no "middle way" between the two. By thinking through each scenario, suspending for a time the urge to choose one and "predict", his book became a powerful tool for facing the implications of the choices available to South Africans in the mid-1980s. Sunter's book became a minor classic; even in 1995 I saw it on sale in a South African news agency chain.

Many companies conduct in-house scenario planning exercises. Some of these are kept confidential; others are published, in full or in part. Two recent Shell scenarios are "New Frontiers" and "Barricades". In "New Frontiers", economic and political liberalisation increases, leading to faster economic growth (and higher demand for energy) but at the same time to more competition. In "Barricades", the world becomes more divided and regulated, economic growth is lower, and business is overcontrolled by governments keen on maintaining national or regional strategic advantages.[4]

Having gathered our own group of experts at the IP, we set out to follow a series of steps to produce the scenarios in Chapter 10 of this book. The remainder of this chapter describes how we did so.

As with any intellectual process, it is best to begin by asking exactly what it is you are doing: defining the question. We set ourselves the broad task of thinking about the oil and gas industries on the UK Continental Shelf in 2010: what would they look like?

One of the first steps is to gather information. In some exercises conducted by companies or consultants, this is a lengthy process. We chose instead to assemble a team of participants who had senior and wide-ranging experience of all aspects of our subject, and who had the reputation within the industry of having original minds; they reflected the wide range of membership of the Institute of Petroleum. They were all also potential users of the workshop's product: they are all engaged in decision making which affects the future of the UK's oil and gas industries. (The participants are listed in the introduction to this book.)

We chose the participants carefully to cover a range of interests and approaches: representatives of the "major" international oil and gas companies, and of the "independents"; of the supply industries, finance, politics, the downstream "customers", and so on. This diversity is a luxury which it is difficult for one company, engaged in its in-house planning, to provide. Secondly, within any one organisation there is always the possibility of "groupthink"; the unconscious reliance on shared assumptions and languages, of wishful thinking or internal political correctness. Members of our group came with widely differing perceptions of the offshore industry, and indeed of the usefulness and methodology of scenario planning itself.

Given this base of knowledge and experience, we then attempted to identify the main factors which will influence the fate of the industry over the next 15 years. To put it another way, if we could by some magic know the answer to two or three questions about the future, what are the most important questions we would choose to ask?

Peter Schwarz, in his book *The Art of the Long View*,[5] stresses the importance of identifying, and ranking, the "drivers" that will influence the future. "Drivers" may take many shapes. They may in some rare cases be known or reasonably predictable. The best example is the population of a country; barring major disasters such as epidemics or war, or significant changes in national immigration policy, the population of the UK in 15 years' time can be reliably estimated. Most of the people who will be alive then are alive today; death rates and birth rates are known, and are unlikely to change greatly in 15 years. Other drivers are largely unpredictable: the oil price, for instance.

Clem Sunter approaches drivers in a different way: he identifies "rules of the game" and "key uncertainties". All of these are, in our sense, "drivers". The advantage in articulating the "rules of the game" is that you become aware of the assumptions that you are making. One of Sunter's rules, for instance, is that violence in South Africa would rise under a bad political scenario. Military and police power could not completely control expressions of popular discontent. One of his "key uncertainties" was the strategies that would be adopted by the various political groups, summed up in effect as negotiation or war.

We began our session by "brainstorming" the possible drivers. To do this, we deliberately encouraged each person to think in isolation: to write down on a pad all the drivers which occurred to them, one driver per page. We then grouped the pages on a board. As expected, some drivers were chosen by more than one person. Often, however, each driver had been described in slightly different ways or from a different angle. As we grouped them, it also became obvious that they were inter-related.

We reduced these drivers as much as possible by eliminating duplicates and combining those that were closely connected. We had a little time to examine each of these, although not in detail, and to look at the sort of limits within which it might vary: not to predict what it would be like in 2010, but to establish plausible ranges within which we might expect it to lie. The drivers that we fixed on, and the limits that we envisaged, were as follows.

THE FIRST DRIVER: NEW TECHNOLOGY

We included such developments as subsea engineering, 3D and 4D seismic, and deep water operations. Favourable technological developments would include cheaper floating production storage and offtake vessels (FPSOs), subsea multiphase pumping and metering, and "off-the-shelf" designs for platforms and other hardware. Engineering would tend to simplicity, reliability and standardisation. In a favourable world, technology would introduce a wide range of possibilities, including the ability to produce and recover over long distances without the need for platforms, and better exploration tools that could improve finding rates for new fields, and the ability to analyse and manage existing reservoirs to give recovery rates of 70% or more. In an unfavourable situation, research and development would either be unsuccessful or unfocused: it would fail to deliver cost-effective new solutions to existing problems, or to widen the technical "window of possibility".

THE SECOND DRIVER: THE COSTS OF EXPLORATION AND PRODUCTION

How far can these be cut? If all went well, we assumed considerable cost reductions, so that (for example) 10 000-foot wildcat wells could be drilled for $2 million, and that stand-alone fields of 5/10 mboe would become economically viable. If not, then the industry might return to the risks of "gold plating": costs might actually rise from present levels, and returns fall. In this situation, if the oil/gas price was high, the companies would continue to operate; but if the price stayed low, all but a few developments would become uneconomic.

THE THIRD DRIVER: THE DEVELOPMENT OF RELATIONSHIPS WITHIN THE INDUSTRY

At best, there would be streamlined and synergistic relationships between operators, suppliers and other contractors and subcontractors. They would be well managed and driven by agreed common goals. In an unfavourable future, these relationships would be combative and inefficient: each company would seek its own interest, whatever it might claim to be doing, and overall management would be divisive and unable to control the process properly.

THE FOURTH DRIVER: THE INDUSTRY'S OWN INTERNAL ORGANISATION

This includes the trend towards fewer people, with more skills, and the ability of organisations to learn. In a favourable future, the industry would adapt its own structure to meet the business environment and challenges it had to face. At worst, the industry would fail to adapt. In a high oil price environment, it might succumb to the temptations of "gold plating" – too many people engaged in non-essential activities. Or it might become over-bureaucratic: perhaps in response to external pressures or in an attempt to maintain perceived quality, it would concentrate on bureaucratic "paper" controls at the expense of real safety, quality and environmental effectiveness.

THE FIFTH DRIVER: THE OIL AND GAS PRICE

This includes the external factors that will influence prices, such as world demand and events in other producing areas. In a favourable future, we assumed high prices in 2010 of $20 per barrel/20 pence per therm (in 1995 money terms). These would be stable price levels, with no recent or anticipated fluctuations of any size. At worst, we assumed a low oil price of $10 per barrel/8 pence per therm, or a price which might fluctuate from day to day but was not be expected to rise above this level in the medium or longer term.

THE SIXTH DRIVER: THE DEVELOPMENT OF THE GAS MARKET

This driver includes the growing importance of gas in the offshore mix, the European gas market, and gas pricing practices. In a favourable future, there would be a increased demand for gas as the UK and mainland European markets expand, and/or gas captures a greater share. At worst, there could be dumping of imported gas, continuing price collapse, and chaotic fluctuations in the supply/demand balance.

THE SEVENTH DRIVER: THE AVAILABILITY OF MONEY AND THE DEVELOPMENT OF RISK MANAGEMENT TECHNIQUES

In a favourable future, capital would continue to be available for projects in UK waters with a competitive rate of return. More risk management tools would be developed to help borrowers and lenders structure their risk portfolios. At worst, a shortage of capital and a perception of high risks might curtail all but the most obviously profitable developments. Other geographical areas or industry sectors would appear more attractive to lenders. Risk-management tools would lose credibility.

THE EIGHTH DRIVER: THE EXTENT OF BRITAIN'S UNTAPPED OIL AND GAS RESERVES

Where will any new reserves be found, and how large will they be? In a favourable future, new fields of reasonable size would continue to be found in frontier areas, and smaller fields developed in the areas which have already been well explored. Licence acreage would be fully exploited. At worst, the new frontier areas would be disappointing. Companies holding licences would sit on smaller fields and prevent them being developed by others.

THE NINTH DRIVER: THE AVAILABILITY OF OFFSHORE INFRASTRUCTURE

This includes both technical availability, and the commercial terms on which existing infrastructure will be available to new users. In a favourable future, the existing infrastructure would continue to serve in new capacities. There would be no technical failures, and companies would make infrastructure commercially available to each other, enabling the development of fields and projects which would not be viable otherwise. At worst, the infrastructure would be prematurely abandoned, or there would be serious failures; infrastructure owners would make it difficult or unattractive for outsiders to share. This would seriously limit the number of satellite fields, or small fields in the main North Sea areas, which could be developed.

THE TENTH DRIVER: THE ACTIONS OF THE UK GOVERNMENT

This driver includes tax rates and licensing policies. Here we assumed that the "best" option is for continuity of existing policies on tax and licensing, with any changes being gradual and with their impact on the industry well thought through. The worst option would be sudden changes, with unexpected effects, leading to uncertainty and reducing the ability of companies to plan ahead.

THE ELEVENTH DRIVER: THE INFLUENCE OF THE EUROPEAN UNION

We included European Union rulings on competition, integration, and directives on working hours and the Social Chapter. In a favourable situation, we saw that there would be little extension of these and similar measures. In an unfavourable world, we saw new requirements and rulings leading to increased costs, limiting managerial flexibility and imposing new bureaucratic requirements. We were conscious that much would depend on the industry's own reputation, but also on the ability of policy makers to consider fully the likely impact of policy measures and the economic implications they will have.

THE TWELFTH DRIVER: ENVIRONMENTAL ISSUES

This includes the impact of any serious incidents that may occur, the development of social attitudes, and the impact of existing and new regulations. In a favourable future, the industry would continue to pay serious attention to its environmental responsibilities. A social consensus would develop that allowed all concerned to agree on the general level of these responsibilities. Industry would be seen to be pulling its weight. At worst, there would be division over what was and was not acceptable, making it impossible for industry to meet agreed standards. Either industry would become overcautious (placing too much emphasis on environmental issues) or some companies would cease to take these issues seriously and would do no more than they can get away with. A major environmental disaster might occur; public reactions would lead to tighter controls and more suspicion, which would limit investment and flexibility. New fuels, made more economically viable by environmental pressures, might take over from hydrocarbon fuels in some areas.

THE THIRTEENTH DRIVER: THE ABANDONMENT OF OFFSHORE STRUCTURES

How can this be done, what will the costs be, and what will the environmental implications be? In a favourable future, we believed, abandonment decisions would not be rushed; alternative uses of the infrastructure would be considered. A consensus would develop between industry and the public/government on how any abandonment that does take place was to be handled in an environmentally optimal fashion. At worst, there would be premature abandonment, before the benefits of the existing infrastructure (e.g. for the development of nearby fields) had been fully appreciated. The abandonment process would be politically controversial and expensive and would consume disproportionate amounts of management time and company resources; it would distort public and government perceptions of the industry.

THE FOURTEENTH DRIVER: THE REPUTATION OF THE INDUSTRY

The oil and gas industries are both more glamorous and more suspect than most other industrial activities. In a favourable future, the industry would be seen as socially responsible. It would attract bright young people to work in it. It would earn the trust of the public and regulators to manage its own affairs well and conscientiously – government and others would "listen" to industry and give its requirements a fair hearing. In the less favourable situation, it could suffer from a lack of new talent, and attract tighter regulations and public suspicion or hostility. This driver depends very much on the industry's own efforts.

The implications of these drivers are examined in more depth in Chapters 6 to 9, which examine some current developments in each area and discuss ways in which these may affect the future.

Our next step was to look at the way in which each of these drivers would affect the others. For example, if the oil price is high, then there is more money available for research and more incentive to develop new ways of reaching oil and gas which were previously not worth extracting. So a high oil price might be thought to lead to more innovation and technical development. However, a low oil price increases the pressure to find new ways of saving cost; it may encourage a radical reassessment of techniques. So a low oil price may also lead to innovation and development. Many other drivers had similar complex relationships with each other. Internal company organisation would depend partly on profitability, but also on technological development. Profitability would depend partly on tax rates, and these in turn might be influenced by the industry's environmental performance on the one hand, and the oil price on the other. Exploring some new areas of UK waters, such as the south coast and the Welsh coast, might be done at a lower cost than others such as the deeper waters west of Shetland; but the environmental risks need more careful handling. On the other hand, ultimate

114 *Waves of Fortune*

disposal of a large deepwater platform provides another set of environmental challenges. In turn, society's attitude to environmental risks might depend on the price and availability of oil.

It is not possible to find some all-embracing equation which would relate all the drivers we selected. Nor is it possible to draw graphs in multidimensional space to show how a change in any one might affect all the others. It seemed as though we had raised more questions than we could possibly answer.

This was the point at which we found the most difficulty in continuing with our workshop. There appeared to be no way of condensing the factors we had identified into any useful scenario form. While a list of questions had some use, it was not the set of alternative images of the future which we had tasked ourselves to produce. We were aware that scenario planning exercises undertaken by other bodies (such as Shell) might take days or even weeks to produce what we were attempting to synthesise in one day. Our greatest strength – the impressive team we had assembled – was also our source of weakness, since it was impossible to get these busy people to take even more time out of their schedules. It would clearly be of little use if we presented IP members with a long list of "if this, then that", and there seemed no obvious order emerging from our chaos.

Eventually, however – a break for lunch helped – we saw that we could group the main factors into two sets of variables: those over which the industry has some degree of control, and those over which it has little if any control. The first set includes technological development and the implementation of new concepts, the industry's own structures and working methods, its cost control and its reputation. We saw them, assuming any tolerable economic background, as the main influence on activity levels in the UKCS ("industry performance level"). Provided oil can be extracted economically and there are markets, the industry should have the ability to function. (We did not consider extreme economic conditions in which such functioning was no longer possible at all: dramatic price collapse, unavailability of capital because of a financial crash, etc.) The second set of variables includes the world economic situation and in particular the oil price, the attractiveness of oil and gas exploration opportunities elsewhere, and government policy in the UK and Europe, over which UK industry has no control and often little influence ("external circumstances").

These two sets of variables can be set out as in Figure 1. The vertical axis is "high industry performance" vs "low industry performance":

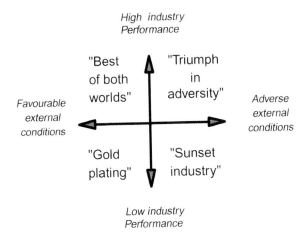

Figure 1 Diagram of scenarios (Source: Institute of Petroleum)

in other words, those factors which are under the control of industry. The horizontal axis is "adverse external circumstances" vs "favourable external circumstances": those factors which are, for the most part, imposed on the industry exogenously. This gives four quadrants, and for each we have produced a scenario: how might the UKCS look in 2010 under this set of conditions?

In the time that remained to us, we discussed each scenario. The scenarios were written up after the workshop finished, and circulated to participants for their comments. Once these comments had been incorporated, the scenarios were published in a limited edition by the Institute of Petroleum. In the introduction to this limited edition, we said:

> Readers may wish to use these scenarios in a variety of ways.
> Firstly, they can be used as test cases for your own planning. (What would we do if any of these scenarios were to happen? How would it affect our own position? Is there anything we can think about now, which will help us to profit from/cope with this future?)
> Secondly, they can be used to help spot trends. Trends are always easier to spot with the advantage of hindsight – and these scenarios help to focus views of the future from which readers can derive a sort of "virtual hindsight". If signs of "gold plating" were to appear over the next few years, for instance – see scenario four – this would lead to conclusions about the forces acting on the industry and the quality of the industry's own response.
> Thirdly and most important of all, the scenarios represent the views of a cross section of the people who currently make the United Kingdom offshore oil and gas industry what it is. They can be read as an invaluable insight into the "movers and shakers" own view of the possible futures for our industry.

Readers may also wish to review these scenarios in five years' time. We do not suggest this out of a masochistic desire to "see if we got it right" – once again, these are not predictions. But if one of these scenarios looks awkwardly familiar in five years' time – or if the industry has taken an entirely different course that we did not consider – then it may be useful to analyse the reasons and ask where rational expectation was right or wrong, and why.

These objectives still seem valid ones, and it is in the hope that the four scenarios will be useful for a wider audience that the present book reprints them.

REFERENCES

1. P Ormerod, *The Death of Economics*, Faber, London, 1994, p 151.
2. S Strange, *Casino Capitalism*, Blackwell, Oxford, 1986, p 25.
3. C Sunter, The World and South Africa in the 1990s, Human and Rousseau/ Tafelberg, Cape Town, 1987.
4. See P Kassler, *Energy for Development*, published in November 1994 by Shell International Petroleum Company Ltd, London.
5. P Schwarz, *The Art of the Long View*, Century Business, London, 1992.

6
Technological Developments

To most British people, George Orwell is a novelist who wrote a terrifying vision of totalitarian future, in his book *1984*. The American-owned oil company Arco British, when it named a southern North Sea gas field the Orwell field, probably did so because of the nearby river Orwell, which flows into the North Sea near Harwich. But in its own way, the Orwell field is a vision of the future worthy of the writer of *1984*. It is a future with minimal human involvement, attractive to accountants but not to romantics.

The Orwell field received government development approval in July 1992. It produced its first gas, in August 1993, after just over a year. Arco claims this is an industry record for fast development of a gas field. All the mechanism is completely submerged below the sea: there is no platform, no floating development facility. Three subsea well-heads resting on the sea bed, 100 feet below the surface, are surrounded by a protective cage some 20 feet high. A 35 km pipeline connects the assembly to Arco's Thames platform, north-east of Great Yarmouth. Running alongside the pipeline is an umbilical, which brings back power, control signals and facilities for chemical injections into the Orwell wells. The Orwell field is actually controlled from Arco's Great Yarmouth shore base, over 100 km away: 120 million standard cubic feet of gas are produced by remote control, without any visible offshore facilities over the actual

118 *Waves of Fortune*

Plate 1 The subsea manifold for the Arco British "Orwell" field, installed in 1993 (Photograph by courtesy of Arco British Ltd)

field. Apart from occasional maintenance, all work on the field is done on land.¹

The giant platforms over Brent, Forties or Ninian, however much they may represent the popular image of the North Sea, are unlikely to be built again on other sites. The Troll concrete platform towed out in 1995 by Shell in the Norwegian sector was widely described as among the last of its kind. Any description of Troll reads as though from another era, when sheer size was a marvel in itself: 430 metres tall, 1.05 million tonnes in weight, expected to last 70 years with a peak production of some 30 billion cubic metres per year. The platform was said to be the largest object ever moved on the face of the earth. Where giant fields such as Troll exist, giant platforms are still a valid option; but even so, technology is changing the face of the industry, making it simpler, cheaper and smaller. The Orwell field is a special case: it is in shallow water, near to existing facilities, and makes intelligent use of available technology. But it points to the future because it shows what may be done more widely in the next 15 years with developed technology.

> *The First Driver: New Technology*
> The Scenario Planning Workshop included such developments as subsea engineering, 3D and 4D seismic, and deepwater operations. Favourable technological developments would include cheaper floating production storage and offtake vessels (FPSOs), subsea multiphase pumping and metering, and "off-the-shelf" designs for platforms and other hardware. Engineering would tend to simplicity, reliability and standardisation. In a favourable world, technology would introduce a wide range of possibilities, including the ability to produce and recover over long distances without the need for platforms, and better exploration tools that could improve finding rates for new fields, and the ability to analyse and manage existing reservoirs to give recovery rates of 70% or more. In an unfavourable situation, research and development would either be unsuccessful or unfocused: it would fail to deliver cost-effective new solutions to existing problems, or to widen the technical "window of possibility".

This chapter looks at some of the technological drivers: at the new capabilities and possibilities that are already on offer, or under development, and which will change the face of the industry. Some are unglamorous: better means of pumping, lighter materials, standardised platform design. Others are as complex as space technology and often similar in concept: new seismic developments, for instance, echo the way that astronomers squeeze more information out of scraps of light from distant stars.

In all these drivers, economics is a key factor. With current low oil and gas prices, technology cannot become an end in itself. Just because a thing can be done does not mean that it is economically desirable to do it. The questions that drive every developing oil and gas technology in the 1990s are: will it help us to discover, drill for, produce and recover oil or gas more cheaply, or more effectively, or more quickly? Will it minimise our risk and maximise our profit, while at the same time meeting, or if possible exceeding, our current safety and environmental standards?

It is vital that research and development remain focused on the critical tasks of oil and gas companies: to explore and produce safely and cost-effectively. "Pure research" can become an expensive sidetrack, and most of the developments discussed here have arisen as means of solving specific technical or commercial problems. Many developments are driven by a need to keep costs down, but have in turn brought about further benefits such as increased safety or less environmental risk.

This section looks at some ways in which technology is meeting these challenges, and attempts to point out future developments.

EXPLORATION AND SEISMIC TECHNOLOGY

Seismic and other exploration technology is often thought of only as a means of finding new fields – the lonely geologist in the unexplored desert. But increasingly it is being used to extend knowledge of already developed areas: to find new pockets of oil or gas in the midst of a known and already exploited field, or to keep track of the way in which hydrocarbons are drained from a known reservoir and to ensure that as much as possible is finally recovered.

This suits the North Sea waters: new large fields are unlikely to be found, but efficient exploitation of smaller ones, which may already be partly known, offers much promise. Allyn Risley, the Chairman and Managing Director of Phillips Petroleum Co UK Ltd, said of his company in 1995 that: "We have a strong acreage position and in recent years have acquired 3D seismic surveys on virtually all our blocks, so we're now in a position to explore with better tools in the coming years. This is one of the reasons I believe exploration in the North Sea will definitely contribute to our growth picture."[2]

A bewildering range of new techniques has been developed to collect and analyse information about events below the earth's surface.

The basic techniques of seismic surveying remain. Air guns towed by ships transmit soundwaves which travel through the rocks below and reflect back to the surface, where they are sensed by detectors. In traditional, "two-dimensional" seismic surveying, only one line of detonations is used in each traverse. In "three-dimensional" surveying, more airguns and more streamers of hydrophones are used, and the relationship between each line is a key factor in the assessment of the results. (In an analogous way our two eyes, slightly to one side of each other, see slightly different images of the same object. Our brains piece the resulting views together to enable us to see in three dimensions and to judge depth and the nearness of objects.)

One of the first uses of 3D seismic was in the North Sea in 1976, by the now defunct Britoil company. However, the technique has only recently begun to come into its own. It has been estimated that 3D seismic increases the chances of an exploration well being drilled in the right place from one in five (with 2D) to one in two. (The actual figures may be more modest: in 1994 Amerada Hess drilled 11 exploration and appraisal wells on the UKCS, and made four major finds. BP drilled 14 but announced only one major discovery.[3])

However, the difficulties and costs of 3D are high. One BP 3D survey of a 2000 sq km area of the West Shetland basin in 1993 cost £13 million.[4] A survey of operators in the Gulf of Mexico found that 57% did not use 3D, partly because it was too expensive.[5] As a result of the high costs, there is a trend towards multi-client surveys, commissioned by several oil companies who share the results. 3D may be used primarily as a development tool on a field which is already producing (and providing cashflow); or it may be used for final analysis of prospects already screened by 2D surveys. However, the costs can be expected to fall.

The development of 3D has been slowed down by the technical difficulties of towing several parallel lines of guns and sensors – see below – but mostly by the immense requirement for computing power to analyse the results. Originally, analysis of a "shoot" might take a year to complete. Now, thanks to the development of much smaller but more powerful computers, and of more accurate means of fixing the position of the survey vessel, the data processing can begin offshore. The overall time from shooting to first interpretation has been dramatically compressed. Amerada Hess, for instance, discovered the Dauntless field (estimated reserves of 50 million barrels) within three months of being awarded the licence in the UK's 15th licensing round; it is claimed that on one well the company mobilised a rig within a fortnight of the (onboard ship) processing of 3D seismic data.[6]

There remains a host of proprietary and in-house interpretation software – there is no industry standard. Although seismic acquisition is done by contractors, seismic interpretation techniques are carefully nurtured "in house". This is an area which oil and gas companies regard as one of their "core competencies". Building platforms or even drilling wells may increasingly be contracted out; but the ability to spot a good "play" and to identify and exploit a reservoir are crucial to success and competitive advantage. Allyn Risley of

Phillips, for instance, said: "I think that if we're not actually ahead in 3D seismic processing and interpretation we're certainly up with the leaders at the front of the pack. And our geophysical specialists believe that we are leading the way in such key areas as post- and pre-stack depth migration, and seismic innovation."[7] (Post- and pre-stack migration techniques compensate for the way that seismic energy is bent by certain geological layers.)

It has been demonstrated that any new improvement in exploration techniques leads to a sudden increase in exploration success, as most of the reservoirs that are exposed by the new technique (but were not visible using older methods) are quickly discovered; and then the rate of finds settles down again. If this is so, then the North Sea can expect a high rate of exploration success over the next few years, particularly in and around existing fields where data has already been acquired and new analytical techniques may enable the companies to read more into it.

Computer processing of seismic echoes is only one step. Geologists need to refine their understanding of the ways in which hydrocarbons are generated, how they move under the earth's surface, and why they are trapped in what subsequently become oil or gas fields. The computer-assisted interpretation of seismic and other data can produce highly sophisticated maps of underground features, but these still need to be interpreted. The movements of the earth can be simulated by sand models; they can also be simulated by much more elaborate geological modelling software packages such as Project Hedera, developed by the Petroleum Science and Technology Institute with funding from industry and the Offshore Supplies Office.

The geologists are now going one step further. Four-dimensional (4D) seismic is a further refinement of the technique. This is a "time-lapse" method, where seismic surveys are repeated at regular intervals in the same place. It is most appropriate for fields which are already discovered and producing. In geological terms, the time intervals between surveys are tiny. Oil- or gas-bearing strata formed over millions of years. The only changes over periods of at most a few years between the successive shots in a 4D programme are owing to human intervention, and particularly to the changes in the reservoirs caused by the production of oil and gas. 4D therefore offers the geologist an opportunity to watch the reservoir as it is drained, noting migration of water into the reservoir, changes in gas–oil ratios, and changes in pressures. This can be invaluable in helping the reservoir engineers to optimise the production profile: it can help to determine whether artificial methods of recovery are necessary, and if so where and when water injection or other types of wells should be drilled.[8] It can also spot pockets of oil or gas which are not being produced, perhaps because they are not connected to the main reservoir; these can then be tapped by small separate wells.

Perhaps the ultimate seismic techniques for existing reservoirs are to place the seismic acquisition equipment or geophones on the sea bottom above the field, or downhole in wells penetrating the reservoir itself. The first approach has been developed by the Norwegian state oil company Statoil. Because there is no interference from layers of sea water, this technology claims to be able to distinguish not only potentially oil- or gas-bearing strata, but whether there is gas or water in the pore spaces in the rock.[9] Fixed seismic points, in close contact with the formation they are measuring, are obviously ideal for repeated 4D timeslices.

One unusual aspect of 3D and 4D seismic is that they are more expensive to collect onshore than off. Seismic acquisition at sea, where the arrays can be towed behind a vessel, is much quicker and simpler than working on land, especially if permission has to be sought or there are environmental issues.

Several developments are making offshore acquisition more efficient and cheaper. Providing a ship which can tow an array of streamers, preferably in almost any weather conditions, is one challenge. The "ram form", designed by Norwegian Roar Ramde, is a startling new shape for the North Sea. The hull is truncated – as though it was the first part of a much longer vessel, cut off abruptly, leaving a virtually triangular vessel with a wide stern which enables it to tow up to 12 full-length seismic streamers, and provides a larger work area. The shape is also said to be extremely stable and to minimise downtime due to weather. Users of more conventional hulls are developing wing-shaped deflectors to help to space out the lines of streamers towed behind the vessel. Information from the seismic acquisition modules in each streamer now flows to the vessel down fibre-optic cables, for lighter weight and greater accuracy. The amounts of data that can now be recovered in a single sweep are so great that conventional tape media cannot keep up with recording them; new high-density data-recording media are being developed.

Storing the information obtained from increasingly sophisticated and complex exploration techniques is a serious problem. Much early data, collected say in the 1970s, may be on older material, but is still of potential use in investigating possibilities of developing satellite wells near to major fields. Most operators have large stores of valuable data, with the associated problems of maintaining the integrity of the media and the security of the information. A pilot scheme to pool geotechnical data, run by a company called Common Data Access Ltd and supported by many operating companies, allows companies rapid "down the line" access to data without having to store it themselves. Initially the service covers well log data, but if extended could include almost any exploration data.[10]

Several other techniques developed for onshore use may also provide new windows into the subsea geology around the UK. Oil or gas reservoirs consist of quantities of hydrocarbon trapped underground, typically by layers of impervious rock above them. However, it is often the case that small amounts of hydrocarbon manage to seep up through the supposedly impenetrable rock and reach the surface. Here, they, or the effects and reactions they have on other components of the earth, can be detected by a range of chemical, magnetic and radiographic techniques. It is even possible to analyse the subsurface soil for specific types of microbial organisms which indicate the presence of light hydrocarbons.[11]

NEW PRODUCTION TECHNOLOGIES

Subsea Completion

This chapter began with a description of the Orwell field, a good example of a subsea completion. Offshore oil and gas wells traditionally have their well-heads above the surface of the sea, on a fixed platform or floating vessel. This is connected by a vertical pipeline or "riser" to the well as it emerges from the sea bed. Subsea completion, or putting the well-head on the sea bottom, has a variety of advantages. It is cheaper: there is no need for a large platform to support the well-head, and all the construction, operation and maintenance costs that go with it. A subsea well-head is, almost by definition, unmanned, and therefore requires less support than a manned platform. It is also safer; there is no riser to be vulnerable to impact or corrosion, and there is no one living on the platform if anything does go wrong. Subsea completion is an ideal technology for the North

Sea, where 96 subsea wells were completed in 1995, and 106 were expected in 1996. The UK represents nearly 40% of the world market for subsea technology, with investments of over $2.8 billion.[12]

The main problems of subsea completion include the difficulty of maintenance, and the need for remote measurement and control. Even minor repairs will require divers or a specialised submarine vessel called a remotely operated vehicle (ROV). Minor repairs will be disproportionately expensive. Both regular and unscheduled maintenance sessions must be few and far between if the installation is to be cost-effective. The system also requires reliable instrumentation and control technology, since it has to be operated from a distance by operators who have to rely on telemetry and instruments to know what the well is doing. It also needs power, for instance to operate valves. Once installed, the success of any subsea equipment is highly dependent on these "intervention costs". The deeper the water in which the equipment is installed, the higher the capital and operating expenditures are likely to be.

A typical subsea assembly is connected to "umbilicals" which carry control signals and its power supplies from a nearby platform. The subsea well produces into a flowline or pipeline. Incoming flowlines may also supply gas or water for injection into a well. The assembly will include a "Christmas tree", the collection of valves which sits immediately on the well-head and controls the flow of the well. It is important to minimise the height of the subsea assemblies, but also to make it easy for an ROV to work on the well. The joints and valves must seal under high pressure under harsh conditions, and also in compact packages. Systems are now rated for work up to 15 000 psi.

In economic terms, the main advantage of subsea systems is that their low cost makes many marginal developments possible. In many cases several well-heads are completed below the surface, and "tied back" to a platform. Sometimes, a whole field may be completed subsea. Equipment is becoming smaller and lighter. Costs are steadily falling: the estimated cost of a subsea installation per well was $20 million in 1990, but only $6.5 million in 1995. For instance, the Phillips Petroleum Company's "Dawn" field, a gas field in the Southern basin, is tied back through a 4.5 mile pipeline to a platform in the Hewett field. This has enabled the field to be developed at a total cost of £16.2 million.[13]

Amerada Hess's Hudson field, one of the most northerly in the UKCS, is another example of the use of subsea technology. It came on stream ahead of schedule and under budget – £134 million against

£190 million. This amounts to an estimated capital cost per barrel of oil of £1.55, plus operating expenditure per barrel of just under £4, excluding provision for abandonment. In this field, first production was through a temporary system, and preceded traditional production through a fixed platform. Two subsea wells were drilled, completed and produced through the floating production test vessel Petrojarl 1; the revenue earned from producing 17 million barrels of oil this way allowed Amerada Hess to develop five more producing wells and two water-injection wells, make the subsea completions, and tie back the whole assembly through an 11 km pipeline to the Shell Tern platform. The Hudson field is a good example of the intelligent and cost-effective use of what is now a rapidly maturing technology.[14]

As well as their cheapness, subsea completions also offer new possibilities for developing deepwater wells; the industry is now able to achieve systems in up to 2000 metres of water. BP estimates that some 30% of the development costs of its deepwater Foinaven field, west of Shetland, will go on subsea equipment and control systems, compared with only 5% in its Andrew/Cyrus project.

At the moment, water depth is a major limiting factor on subsea technology. Not only the capital costs of installation, but also the operating costs such as maintenance, increase rapidly with depth. Given the present rate of development, what may be judged a "conventional" technology for depths of up to 1000 feet can be expected to be available for depths of 3000 feet or more within the next few years. The "Deepstar" project, originally conceived by Texaco in 1992, now includes 15 companies who are pooling their experience and expertise to provide cheaper and better deepwater production technology for the Gulf of Mexico. According to Dr George Vance of Mobil, operations at depths of 6000 to 8000 feet will be possible in the next 10 years. Although deeper water increases costs, Dr Vance estimated that going from 1000 to 4000 feet would only increase costs by 70% to 80%. It is now possible to bring deepwater projects on stream, admittedly in Gulf of Mexico conditions, for as little as $6 per barrel.[15] Platforms are now being installed in the Gulf of Mexico in over 700 feet of water, although it is unlikely that deep fixed conventional structures will be the preferred way of exploiting future deepwater fields.

Brazilian waters also include several deepwater developments, and Brazil's Petrobras is widely regarded as an industry leader in subsea and deepwater technology. BP and Statoil are also active in this field. The European Union has supported, through the THERMIE research

Technological Developments 127

programme, the development of a diverless subsea pipeline connection system and an improved remote operated vehicle (ROV) system.[16] Economic exploitation of the west of Shetlands areas of the UKCS may depend heavily on such improvements in deepwater technology.

Subsea functions are not limited to well-heads. Petrobras has developed a subsea oil/gas separation system. Separating oil and gas allows them to be piped over greater distances, allowing the well to be serviced from a platform in shallower water, or even from the shore. Subsea booster pumps allow pressure to be maintained in single or multiphase flow, to drive fluids over longer distances.

In attempts to drive subsea costs down, various alliances have been set up between companies: either between suppliers of different types of equipment suppliers (ABB Vetco Gray) or between manufacturers of subsea systems, and the contractors who install and maintain them (Kvaerner/Stolt Comex Seaway). Standardisation allows equipment to be more cheaply made, but also permits operators to use the same sets of remotely operated tools, and techniques, on different parts of a field. This makes maintenance easier and cheaper. Subsea equipment can also be designed to be recoverable and reusable. This is partly a matter of designing the components for easy removal and reinsertion by an ROV or by divers, and using standard approaches wherever possible. It also demands the use of higher specification materials to overcome problems with corrosion in equipment, which may be designed for a working life of 30 years. Assessments have suggested that the higher cost of installing recoverable equipment on the first well would be more than compensated when the equipment is used for the second or third time.

Subsea technology offers the prospect of developing new fields without expensive surface structures, and producing the oil or gas over considerable distances back to either an existing platform or directly to an onshore terminal. Given acceptable reliability of the technology, and with capital expenditure costs reduced by reusable standardised equipment, it is likely that the Orwell field will become the rule rather than the exception for small projects close to existing infrastructure.

Smaller Topsides and Innovative Platform Construction

Where subsea completion is not desirable or possible, and water depths are not too great, it may be necessary to build a new platform.

Plate 2 "Mass-produced" offshore platforms: Amoco's Davy and Bessemer platforms (Photograph by courtesy of Charles Hodge Photography)

(Or, in some cases, to take the legs or "jacket" of an existing platform, and put on a new "topsides".) Operators are now reducing the size, weight and cost of platforms, using improved designs and methods of construction and new materials.

Amoco's Davy and Bessemer platforms, in the southern North Sea, are estimated to be about one-third of the weight of an equivalent platform 10 years ago. These are "minimum facility" platforms. They use a "monotower" design – i.e. they have only one leg, supported by struts and a frame on the sea bed. Drilling and well workovers can be undertaken through the centre of the monotower, but the most important advantages of the design are cheapness and the way that its single leg makes it easier to adapt for reuse in different water depths (by adding or taking away lengths of the cylindrical tower: not so easy in a more complex structure). The platforms are expected to be in operation on their current sites for seven to eight years, and then to be reused elsewhere. The two platforms are largely identical; they are designed to Amoco's Minimum Offshore Supporting Structure (AMOSS) outline. The two platforms are estimated to have cost some £84 million, but Amoco has said that if four AMOSS platforms were built at the same time, the unit costs, owing to economies of scale, would come down to three-quarters of this sum. (Or as Amoco put it, the fourth platform would effectively be free.) Several possible uses for future AMOSS platforms have already been identified, and Amoco is talking to other operators with a view to agreeing a standard design.[17]

There are other cost savings from ingenious but minimal facilities on the topsides of the platform. As an example, wind turbines are used to generate operating power: this not only saves on fuel costs, but reduces the need for maintenance visits to refuel and check conventional generating systems. The AMOSS platforms are, like many others, classed as not normally manned installations. That is, they are not intended (or licensed) for regular human occupation: only short maintenance visits of a few hours are envisaged.

While the conditions of the southern basin are relatively easy and favour smaller and simpler platforms, other operators have developed similar ideas elsewhere. In the Netherlands sector, for instance, Clyde is working on a reusable production facility and Wintershall has installed three platforms which, although not identical, are built to common designs wherever possible. This allowed them to be built almost on a "production line", with time and cost reductions.[18]

New materials are also being tested for use in platform construction. Glass reinforced plastic (GRP) and fibre reinforced plastic (FRP), for instance, are now under study by a UKOOA workgroup. These materials have been used offshore for some years: they are often cheaper, lighter and more resistant to corrosion than steels. However, their use in the North Sea has so far been held up by fears that they may be a fire risk – either flammable in themselves, or liable to collapse when heated or to give off toxic gases. A feasibility study for an "all composites" topside structure is under consideration; in the meantime, these materials are used for applications such as piping (e.g. for tubing or liners in seawater injection wells, or for fire extinguishing "deluge" systems on platforms). By the turn of the century, it is expected that they may be used for blast and fire protection panels, module panels, subsea flowlines, and subsea protection structures. (Shell already claims to have saved some $500 000 by using GRP sandwich, instead of steel, for well-protection structures in the Norwegian Draugen field.) US studies are also looking at the use of new materials for coiled tubing and risers.

Another "new" material being considered is titanium. This offers lighter weight and longer life, although initial costs are higher than those for steel. Conoco has used titanium for the drilling riser in the Norwegian Heidrun field, with a 75% weight saving. Instead of pure titanium, producers are experimenting with high-strength steels combined with thermally sprayed titanium, or high-alloyed steels ("duplex" or "super duplex" steel) which show weight savings over traditional carbon steel, as well as lower life-cycle costs through reduced maintenance and less frequent replacement. Aluminium has also been used: Phillips Petroleum estimates savings of 335 tonnes in weight and £900 000 in cost by using aluminium decks on the Judy/Joanne platform. Improved types of concrete are also possible new construction materials.[19]

These developments offer companies a wide range of options for saving capital expenditure and operating expenditure costs on new platforms. They also suggest that platforms will be capable of longer life, at a time when new fields are expected to be smaller and therefore have shorter lives. This combination is likely to bring even more pressure to bear on operators to reuse platforms and other structures: either by moving whole platforms, or by taking existing structures and converting the topsides.

The length of life of platforms can be further extended by new techniques for optimising maintenance. Traditional maintenance uses a combination of two methods: regular inspection of components to detect potential failures and replace them, and anticipation of the expected life of a component, replacing it as its expected life expires, regardless of the condition it is in. With limited access platforms (either subsea well-heads or "not normally manned" platforms), both regular inspection and pre-emptive replacement become more critically expensive. Projects such as RISC (Reliability based Inspection Scheduling) developed by Technical Software Consultants and the University of London and supported by the EU's THERMIE programme, have attempted to pull together and optimise the various aspects of maintenance planning. The techniques used include better databases of component reliability, and improved modelling of the effects of stress and deterioration. The programme can also optimise the scheduling of maintenance, minimising downtime and the use of resources. While RISC is particularly aimed at fatigue in tubular joints, other systems have been developed which use broadly similar techniques to optimise maintenance schedules. Although these systems are not technologically glamorous, they may save considerable amounts of money as well as increasing safety and reliability.[20]

DRILLING TECHNOLOGY

Traditional drilling has been done with lengths of rigid tube, "drill pipe", produced and handled in 10-metre lengths, which are screwed together to make up the drill string. The need to provide a framework to raise these lengths of pipe above the well while they are screwed together and lowered one by one has given the drilling rig its familiar shape. Most drilling is still done in this way.

One radical alternative, coiled tubing (CT) drilling, replaces the rigid pipes by a flexible tube. This offers many advantages: ease of handling, smaller, lighter and cheaper drilling rigs. The initial constraint on using CT has been the limited size of tube that can be produced. Until the late 1980s, CT was available only in very narrow gauges, up to 1 or 2 inches outside diameter or OD. (The OD of conventional drill pipe used in a typical well is typically between 15 inches near the surface and 9 or 7 inches at the bottom of the hole; risers may be up to 30 inches wide.) Consequently, since it was introduced in the 1960s, coiled tubing has been mostly used for "workovers" and "well servicing" – maintaining or developing existing wells. For instance, the BP Magnus field, which made no use of CT in 1986, found 15 uses of the technique in 1993 for such purposes as cleaning out existing wells and "fishing".[21] Cleaning out wells may become necessary if sand, chemicals or other deposits form in them during production, clogging them up. CT and the use of new milling or jetting tools enable a controlled clean-out. "Fishing" is necessary when objects such as downhole tools have been accidentally dropped into the well, or have become stuck. CT allows the operator to remove debris while keeping the well under control.

The technique is particularly useful for extending existing wells, for instance by drilling new horizontal lateral arms out from existing holes. Where a well has been in production for some time and flow is falling off, this technique may enable it to tap new reservoirs (or new parts of the same reservoir) without any additional construction above the surface.

CT has seen a resurgence in the 1990s. As the available size of tubing becomes larger, the technique is being considered for drilling new wells. Advantages of CT drilling include greater safety: there is no need to connect and disconnect lengths of drill pipe, so there is minimum exposure of personnel on the drill floor to the risks of handling and spinning heavy items. The coiled tubing string can be maintained under pressure even while it is being pulled out of the well, so that mud circulation can continue and primary well control is maintained. The CT unit is much smaller than a conventional drill rig: for this reason many early uses have been on land, where the units are less of an eyesore and cause less environmental damage. Offshore this translates into smaller platforms.

The increased use of CT has coincided with new interest in drilling narrower wells. So-called slimhole wells, typically with a final diameter of under 5 inches, are considerably cheaper to drill. Using traditional methods of drilling, the hole must be tapered, starting with wide casing sections at the top and gradually narrowing down as the well is drilled. This is not necessary with CT. CT and slimhole drilling (SHD) may be most often used on land, where the advantages of minimal disturbance, a small site and speed and cheapness are most likely to appear. However, the technique also offers drastically reduced costs for exploitation of small offshore fields. It minimises the need for mud, cement and transport to the rig. It has been estimated that a slimhole well, with a top section of 13.375 inches and a lower section as small as 3.5 inches, may cost only 60% as much as a conventional well (24.5-inch top section, 8.5 inches at the bottom) and produce only 30% of the volume of cuttings.[22]

The chemistry of mud, or drilling fluid, is another area in which significant developments are occurring. Traditional mud uses oil as a base to circulate the chemicals used to "weight up" the mud, to balance the pressure of oil or gas in the well. When mud is circulated it returns to the surface, bringing the "cuttings", or pieces of rock which have been drilled out of the hole. These are usually dumped over the side. They are, however, contaminated with the oil from oil-based mud, if such a mud has been used. Current regulations require the oil to be cleaned off to protect the environment before the cuttings can be dumped. As a result, operators often now use mud based on water, or on biodegradable "pseudo-oils", to minimise the need to clean cuttings. (Wells can also be drilled using dry air or gas at high pressure, or a mixture of air and mud.)

However, drilling mud performs several functions. It has an important role in ensuring that the sides of the well do not collapse. For this reason, some types of formation can still only be drilled with oil-based muds. As mud chemistry develops, this constraint may disappear. Mud and downhole fluids are now very precisely engineered to cope with the different problems that are expected as a well is drilled. They may be formulated to kill bacteria which live in the oil, or to reduce calcium deposits inside the well, or to inhibit corrosion. Special muds may be used at precise points in the drilling programme to reduce expected friction, or to control chemical reactions between the drilling fluids and certain types of rock strata.

The material from which the drill strings are made is also being developed. Current traditional drill pipe is heavy and relatively inflexible. This can lead to strain and even sticking, especially during horizontal drilling. Alternative materials such as aluminium, titanium or composites have been considered. If these can be provided at competitive prices, they have many advantages: composites, for instance, may have much greater resistance to chemicals and corrosion, and any lighter material reduces the loads on the drilling rig and the amount of energy used.[23]

Drill bits are another area where there is a lot of research and improvement, to increase both the life of the bit and the rate of penetration (ROP). Improvements mean that the well can be drilled more quickly, with less delay while the bit is removed and changed. This procedure, in a traditional well, involves taking out the entire drill string, unscrewing it piece by piece, until the bit is out of the hole. The bit is then changed and the drill string laboriously lowered back into the hole, screwing each piece of drill string back into place as the previous one is lowered down by the derrick. This is known as a "round trip", or "tripping". In a well that may be a mile or more deep it can take a day. Anything which increases bit life, and the life and reliability of the other components at the sharp end of the drill string, brings about an immediate saving in time. "Tripping" is also relatively dangerous for the drilling crew, since handling the drill pipes involves skill and risk. Drill bit manufacturers are introducing a wide range of proprietary technologies to make bits more reliable and effective; these include use of materials science to produce harder and more durable cutting surfaces, and better design to minimise vibration and to ensure the cutting surfaces contact the rock formation at the optimal angle.[24]

The shape of the wells themselves is also changing. Where once they went straight down, they may now be drilled to curve (or "deviate") away, enabling one fixed platform to put wells down that end up several miles apart. This technology has been available for some years and is regularly used; however, it is developing and extending both the range of the wells and the precision with which they can be "steered" into the precise position where the reservoir is expected to be.

Deviated wells are often drilled without the traditional "top drive": the bit cannot be efficiently turned at the bottom of the hole by a motor on the surface rotating a whole rigid length of drill string. Instead, the bit is turned using a "mud motor", which is installed at the bottom of the drill string and powered by mud being forced down the hole, through a turbine in the mud motor, and back up the hole again. This constant flow of mud turns the motor and provides motive power for the drill bit; the string itself no longer needs to turn, thus reducing possible friction. Mud motors are, however, expensive. Reliability is important: there is no cost saving if they have to be returned to the surface for repair or replacement more often than the drill bit itself needs to be changed. Improved seals, bearings and other parts of the motor are critical. The motors also require careful handling: although they can operate with many different types of mud, including very heavy muds used to control a well, they can easily be brought to a halt if the mud is not properly mixed or if there are abrasive solids circulating in the well. High temperatures at the bottom of the well also pose problems. However, the mud motors make new shapes and lengths of well possible.

Wells drilled in the Norwegian Gullfaks field have reached out horizontally more than 4000 metres, including turns in two directions.[25] Esso expects extended reach drilling to reach out to around 9000 metres by the end of the century.[26] Techniques for monitoring and controlling the drilling are being refined, to enable the bit to be placed exactly where it is needed.

The original use of directional drilling was to reach out to greater distances from existing platforms, so maximising the area that one platform could tap. The technique is now also used to control the angle at which wells approach and enter reservoirs. Oil- and gas-bearing reservoirs may in some cases be quite narrow strata of rock; if they can be threaded through, like pushing a syringe into and along a vein, more of the contents can be extracted than if the well simply cuts vertically through the vein in one place. Directional techniques enable smaller reservoirs to be reached without the need for expensive new structures above the surface; they also enable smaller reservoirs to be more completely drained. They even enable the driller to steer round other wells, or difficult structures, on the way down.

Planning such a well is very complex and requires computer simulations. There are limits on the tightness of the bend that a well can be made to take, which in turn depend on the thickness and flexibility of the drill string used at each point, and the length and complexity of the equipment involved. Applying weight to the drill bit is an entirely different prospect when the well is partly horizontal; the frictions involved are also much more difficult to anticipate. Making the best use of geological structures requires an exact and detailed map of those structures, and very precise manoeuvring. A "designer well" of this sort is an impressive engineering feat: highly knowledge or skill intensive, even though the actual equipment may be smaller and lighter than its old fashioned counterparts alongside on the same platform.

Once planned, the well must be drilled accurately. Downhole sensors give the driller a variety of information, for instance about the angle of the drill bit or about the resistivity of the formation and other factors, enabling him to steer the bit with greater precision. Increasingly this information is available in real time, or within minutes of the measurement being taken (the so-called MWD, or measurement while drilling). Well-logging techniques, which give the driller information about the structures through which he is going, are also increasing in sophistication and help to make best use of the driller's new flexibility.

PRODUCTION TECHNOLOGY

Finding the reservoir and drilling the well, however technologically challenging both may be, are the beginning, not the end. They must be done cheaply and accurately. But the objective is to produce oil or gas. This is a less glamorous process, but one in which there are also continuous developments.

Production may extend over many years, so any cost savings per unit produced will add up over time and have major economic implications for the field. Production includes a range of processes: controlling the flow of the producing well, separating different produced fluids (such as oil, gas, condensates and water), performing some initial processing, and recovering them to shore, either by pipelines or tankers. Processing work done at the well-head is, of course, more expensive, since the equipment has to be taken out to sea rather than operated on land. Such processing as is done is usually only for the purpose of making it easier to recover the fluids to the shore.

The use of lighter and cheaper platforms, or subsea developments, has already been covered. So has the use of existing infrastructure. These are often the cheapest and best sites for initial processing when the new well is sufficiently close to an older facility. When it is not, temporary facilities may be used.

FPSOs

Floating production, storage and offloading vessels (FPSOs or "floaters") are now starting to appear over smaller fields. These vessels replace all the activities normally performed by a fixed platform: they can provide not only production capacity, but also water injection and gas lift. BP will use an FPSO, the "Petrojarl Foinaven", on the Foinaven field. A typical FPSO, the "Uisge Gorm" operated by Bluewater, provides storage for 620 000 barrels of oil. Shell's proposed "floater" for the Teal field will have storage capacity for about 850 000 barrels. This sort of storage capacity allows a field such as Amerada Hess's Hudson, producing at around 30 000 barrels a day, to operate without building a pipeline. Produced oil is stored in the tanker and transferred to shore by a shuttle tanker, visiting as often as necessary.

FPSO designs may either be monohulls or semi-submersible rigs. Semi-submersible rigs, however, lack the required load capacity, and usually do not provide storage. Monohull or "ship shape" designs based on oil tankers are favoured for many new designs, although they cannot provide the same drilling or workover facilities as a semi-submersible. Some new designs, such as McDermott's TMP project, claim to combine the advantages of each type. Other operators have used both a semi-sub rig and a converted tanker, working together.

Technological Developments 137

Plate 3 The FPSO "Petrojarl Foinaven" floats out from drydock in Spain in 1994 (Photograph by courtesy of British Petroleum)

Whatever design is chosen, the FPSO needs the same robustness and ability to withstand bad weather as a drilling rig; in addition, it must be able to transfer large amounts of oil quickly to a shuttle tanker, without risk of environmental damage through oil spills. Mooring systems are critical. These must allow a monohull design to swing like a weathervane to minimise the impact of waves, while guaranteeing leak-tight connections to subsea well-heads. The FPSO must also be able to handle a large throughput of oil: not only must it take the production of a field, but it must also be able to load a tanker quickly. Some experts now believe that a throughput of 200 000 barrels a day is possible, making FPSO technology available to a much wider range of fields.[27]

The FPSO concept is not new. As described in Chapter 3, the Argyll field, operated by Hamilton Oil (now BHP), went on stream on 11 July 1975. Not only was it the world's first floating production system, it was the use of a converted semi-sub rig that allowed Hamilton to steal a march on its competitors and produce the first oil from the

North Sea.[28] (The Argyll field was abandoned in 1993: another advantage of FPSOs is that there are none of the costs involved in abandoning a fixed structure.) MSR's "Emerald" Field has been produced since August 1992 by a converted semi-submersible rig, the "Emerald Producer"; the oil is sent along a short subsea pipeline to a converted tanker, the "Ailsa Craig", where it is stored until collected by a shuttle tanker. (This field is scheduled to be abandoned in early 1996.) The UK's first purpose-built, permanently moored production vessel was brought into production in 1993 by Kerr-McGee over the Gryphon field.

The advantages of using an FPSO are largely economic. As Hamilton found, development time, and hence the time before oil revenues are earned from the field, is shorter. The capital costs are often less than those for a fixed platform, especially if a pipeline must also be laid. The "floater", with a life of perhaps 20 years or more, can produce a small field over a period of five to ten years, and then move on to another field at minimal cost and without the need for the potentially expensive abandonment and removal of a fixed installation.

Because floaters can be reused, contractors such as Bluewater or Golar Nor are prepared to build them and hire them out, in much the same way as drilling rigs are now owned and hired out by contractors. Various commercial arrangements enable the risk to be apportioned between contractor and oil company. This is often the main advantage for the oil company of using a floater: it is chartered, and thus becomes an operating expense, with payments spread out over the life of the field. The heavy up-front capital cost of a fixed platform is avoided. Less capital is required to develop a field, risks are less, and time to production is less. Converting Golar Nor's "Petrojarl 1" floater to move from Amerada Hess's Hudson field to Arco's Blenheim field took only five weeks.[29]

Larger oil companies, such as BP, Shell or Statoil, have the capital resources to build their own floaters. Even when the oil company has to pay the construction costs up front, use of floaters may still be cheaper. Statoil estimated in 1994 that the development costs of the Norne field using an FPSO would be $1.3 billion, compared with $1.6 billion using a fixed platform. Shell's FPSO is intended to develop a group of central North Sea fields. Shell Expro's production director Brian Ward identified the benefits of using an FPSO as "no repeated capital expenditure for conventional fixed platform facilities, shorter lead times for future fields through relocating the FPSO, and, by allowing smaller fields to be developed economically, reduced appraisal drilling requirements".[30] BP is also working on the development of the Foinaven field using floating production concepts.

It is possible that FPSOs will develop in two directions. Those built for specific applications, largely by the major oil companies, may have extensive suites of recovery capabilities. "Generic" designs, built speculatively by non-oil companies, will be capable of operating almost any small field with minimal conversion, even though they may lack the capabilities required to squeeze the last recoverable drop out of every field and some resources may be lost as a result. As a result, the economics of FPSO production may lead to more fields being developed, but lower recovery factors in some of those that are reached.

A Dutch consortium is reportedly working on a scheme to lease removable production facilities, including pipelines, to operators. This would enable a field operator to incur almost no capital expenditure costs: both drilling and production would be carried out by third-party contractors and paid for on a day-rate basis.[31] And, where floaters are not possible or desirable, Clyde Petroleum has developed a platform which can be locked down to a prepared base on the sea bed, and at the end of production unlocked and moved.[32]

REFERENCES

1. *The Orwell Field*, leaflet produced by Arco British Ltd, no date.
2. Interview in *Petroleum Review*, April 1995, p 151.
3. *Offshore*, April 1995, p 113.
4. *Horizon*, BP Exploration, March 1994, p 10.
5. Article by M A Lawrence, H T Logue and D A Grimm, *Offshore*, April 1995, p 26.
6. *Offshore*, April 1995, p 113.
7. Interview in *Petroleum Review*, April 1995, p 152.
8. *Oil and Gas Journal*, 27 March 1995, p 55.
9. *World Oil*, October 1994, p 29.
10. *Euroil*, April 1995, p 10.
11. *Oil and Gas Journal*, 6 June 1994, pp 47–65.
12. *Euroil*, June 95, p 17.
13. *Oil and Gas Journal*, 6 March 1995, p 33.
14. *Offshore*, April 1995, p 42.
15. *Euroil*, September 1994, p 20.
16. *Offshore*, April 1995, p 42.

17. *Oil and Gas Journal*, 27 March 1995, p 30.
18. *Euroil*, April 1994; p 44.
19. *Euroil*, July 1994, p 18.
20. *Euroil*, October 1994, p 33.
21. *World Oil*, June 1994, p 41.
22. M J Pink, Exploration and appraisal technology, Shell Selected Papers, Shell, 1992.
23. *Offshore*, February 1994, p 58.
24. *Offshore*, April 1995, p 48.
25. *World Oil*, June 1994, p 27.
26. *Euroil*, October 1994, p 18.
27. *Euroil*, September 1994, p 18.
28. *Offshore*, May 1994, p 60.
29. *Euroil*, April 1995, p 40.
30. *Oil and Gas Journal*, 23 May 1994, p 25.
31. *Euroil*, April 1994, p 44.
32. *Euroil*, April 1994, p 44.

7
The Way the Offshore Industry is Organised

This chapter looks at a further set of drivers, related to the way in which industry organises itself. For the most part, patterns of organisation are not forced on the industry from outside. The companies making up the industry have a range of cultures and styles. Operators dominate the market; they differ greatly in size, and in the degree to which they subcontract parts of their operations or handle them in house. Drilling companies and other contractors also vary greatly in size, and even more in the size of their UK offices. There is no general agreement about the best business structure for any particular activity.

Instead, there is a continuing debate about the way in which oil companies are organised; their relationships with their suppliers and rivals; the extent to which they can reduce costs without losing their core capabilities; and the management techniques available to them to control their risks. Many of these debates are very relevant in human terms: to the employees of the companies and the communities in which they work, and to the face of the industry.

THE NEED FOR COST REDUCTIONS

The motivation behind cuts in staff is, of course, the reduction of expenditure. Given the collapse in the oil price in the mid-1980s, and

142 Waves of Fortune

the gradual realisation that the price was going to stay low, the need to cut costs has slowly come to dominate the industry.

> *The Second Driver: the Costs of Exploration and Production*
> How far can these be cut? If all went well, we assumed considerable cost reductions, so that (for example) 10 000-foot wildcat wells could be drilled for $2 million, and that stand-alone fields of 5/10 mboe would become economically viable. If not, then the industry might return to the risks of "gold plating": costs might actually rise from present levels, and returns fall. In this situation, if the oil/gas price was high, the companies would continue to operate; but if the price stayed low, all but a few developments would become uneconomic.

The economics of cost reduction have never been more clearly explained than by Dr Rex Gaisford of Amerada Hess in 1994 (Dr Gaisford was a member of the Scenario Planning Workshop).

> For the last 25 years we have built and then believed in, and defended, a high cost culture. We no longer need to believe in the inevitability of high costs. We can now start to believe in and build a low cost structure for the second half of the North Sea story. You, me, all of us need to be a low cost producing area if we are to survive at $13 a barrel oil. There is absolutely no reason why we should not.
> Where does this $13 go? Well, $1 goes in just running the company. That leaves $12. $1.5 to $2.5 goes in the cost of exploring; that leaves $9.5 to $10.5. But exploration expense typically occurs 5 to 10 years before any oil it funds is produced. Even at today's cost of capital, that costs an extra 40% to 50%. That leaves $8.5 to $9.5. We have not started to develop or produce it yet.
> If we take into account the eventual abandonment of the facility and the cost of funding the development itself we are lucky if we have got $7 to $8 per barrel to both develop and produce the field. You cannot do that at traditional prices in the North Sea. Oh, and nobody has made any profit for the shareholders yet.[1]

According to the annual survey by the *Oil and Gas Journal* of the 300 largest publicly traded oil- and gas-producing companies in the USA (the OGJ 300) the average return on shareholder's equity for this group in 1994 was 9.4% (this average has varied between a high of 13.8% in 1982, and a low of 3% in 1986). In order to continue to attract capital, companies have to aim for consistent returns of around this order: this is the target their shareholders have in mind. Oil companies are, of course, very international in their outlook, and they will operate where they can obtain the best rewards. In Dr Gaisford's list of costs, running the company, exploration and capital are much the same wherever the company operates. The difference between profitability and loss is likely to lie in the development and production costs, where the UK's offshore fields have the built-in disadvantages of distance, bad weather and depth. If international capital is to stay in UK waters, production and development costs must be kept down.

This does not mean drastic cuts in activity; rather the reverse. A study by Professor A Kemp and B MacDonald of Aberdeen University suggests that cost savings will actually encourage further field developments.

Reductions in the time taken to bring projects into positive cashflow have significant effects on the NPVs to investors. This is particularly so where projects are only marginally profitable: shorter development times can make all the difference between profit and loss. In effect, they cut interest charges, and transfer money away from the financial community back into the offshore industry.

In an operating environment of capital rationing, reduced costs and faster cycle times enhance the likelihood of obtaining capital, while reducing the amounts required. Kemp and MacDonald see this as a positive sum game which will increase industry activity and cashflow, as well as government revenue, over the next few years.[2]

Elsewhere, Professor Kemp has estimated that if prices stay at $15 per barrel until 2015, many planned projects will not be viable. But, if average savings of 20% in development and operating costs can be made, this would enable the planned projects to be carried out and 27 extra projects to be undertaken. The cost of the extra work which the saving enabled would, at times, exceed the value of the savings made; in other words, the same or more money would be spent, and more oil and gas would be more efficiently recovered for the price.[3]

HOW CAN IMPROVED PURCHASING ENABLE COST REDUCTIONS?

Many companies have attempted to make cost reductions individually. However, there has been a growing feeling that joint action would be more effective, because of the nature of the technical problems involved, and also because the industry structure and methods were becoming overly complicated. In response to this, the Cost

Reduction Initiative for the New Era (or CRINE) was set up in 1993 by a joint working group from industry and government. It has aroused both enthusiasm and suspicion. CRINE is not a standing bureaucracy; a very small secretariat coordinates the work of several committees and working groups. The committee members are drawn largely from industry but the initiative is jointly funded by industry and government. CRINE is not unique; there are similar systems elsewhere in the world, such as the Norwegian NORSOK. CRINE's "competitive advantage", if it has one, is probably that it is voluntary and non-prescriptive.

CRINE has identified and focused on several areas, many of which, like the invention of the wheel, seem blindingly obvious in hindsight, but have taken many years to achieve. Much of its work is to simplify the processes that the oil companies have themselves made more complicated over the years. One target is greater use of standard equipment and "functional specifications". Rather than oversee the design of every part of a platform, it is argued, oil companies should leave the details to their suppliers.

As an example, it is usual for offshore platform cranes to be individually specified. This means that a company building a platform would specify in detail exactly what sort of crane was required, and have it custom built. (In the earlier days of the industry, or when there are special challenges to be faced, this made more sense. It is not possible simply to put an ordinary crane on a platform that must operate in extreme conditions, with limited space.) As an example of more modern "best practice", the CRINE secretariat quotes a case history in which a client/contractor/vendor team discussed requirements and available standard cranes, and decided that a standard, mass-produced ship's crane would be fit for the purpose. This effectively halved the cost.

Documentation and certification are further areas where there is still much duplication. In another "best practice" case study, a purchaser traditionally required certification of each and every component part in high-pressure items (such as valves and manifolds). The manufacturer and client agreed to make do with one certificate for each manufactured item, which referred to the original certificate for each part and allowed each original certificate to be traced if necessary. Saving on documentation acquisition and storage was estimated at £20 000 per year. (Such an initiative, of course, requires the support of government and the Certifying Agency.)

This initiative is supported by a current trend in management thinking – the idea that companies should return to their "core competencies". What business is an oil company actually in? What does it do to add value for its shareholders that cannot be done better and more cheaply by an outside supplier? Lists of core competencies might include the ability to find oil or gas, to appraise finds, and to exploit them in the most economical, or profitable, way; but more and more companies are taking the view that their core business does not lie in designing and building platforms, or (to return to our example) cranes. Nor is it in personally ensuring that each manufactured item has been made from certificated parts. This is the core competency of someone else. An oil company concentrating on its core competencies may prefer to specify its requirements and, where these can be met, to buy the item it needs "off the shelf", more cheaply. If no available item meets its needs, then it will issue a "functional specification" – saying what the crane is to do, but leaving it to the manufacturer to decide how the specification is best met.

Going still further, CRINE is examining the possibility of standardised equipment and working practices, making it more likely that off-the-peg items will be available for any given oil company. Areas in which common working practices are being discussed include structural steel, process and utility pipework, coatings, insulation, and the installation of electrical equipment and instrumentation.

The opposite side of this coin is the need to ensure continued quality from suppliers. If the oil company does not oversee every detail of the production of a component, it must be reassured that the supplier does so competently. The oil company may not include building a platform in its list of core competencies, but if the platform fails after five years of operation, the oil company is inescapably responsible for managing the failure and picking up the threads afterwards. The supplier company which produced the part that caused the failure will not do so; it may no longer exist.

At the moment, companies seek quality assurance from suppliers on an individual basis, each insisting on its own prequalification process to its own standards, and each making its own audits and inspections. This pushes up the costs of tendering and of assessing tenders. Another CRINE initiative is to standardise quality assurance processes into new generic prequalification systems, which all operators could accept – removing the need for each one to repeat the process. This would take away much of the paperwork without lessening the attention to quality and safety requirements.

CRINE has an impact on a very controversial area: that of relationships between the oil companies or "operators" and their suppliers. The operators dominate the market. Once a field is approved and funding available, they can choose from an international range of suppliers. Many suppliers have made heavy capital investments and are vulnerable to price pressures: drilling rig owners are a good example of this. When there are more rigs available than work for them, as in 1994, the owners are rarely in a position to suggest terms of contract to the operators.

There are problems for the larger operators too; they have an in-built tendency to inflexibility. For the best reasons, they may decide to apply worldwide standards for quality and safety; but these may make it almost impossible for local operating units to issue functional specifications. (Previous attempts to standardise working practices have foundered on these rocks.)

As a result, there has been some suspicion of CRINE. Some suppliers regard it cynically as a tool used by some operators to push down prices and offload part of the financial risks, while other operators still retain long-winded methods of specification, audit or purchasing which keep suppliers' costs high. The CRINE Secretariat Director, Vic Tuft, refers to a "cloud of uncertainty" in which the supply industry realises that it is being asked to accept part of the risk, but is uncertain about whether it is getting its fair share of the rewards for doing so. There is no easy answer to this uncertainty; the structure of the industry is changing at the same time as cost-cutting initiatives such as CRINE are taking place. Supplier/operator relationships are at the centre of these changes.

HOW FAR CAN COST CUTTING GO?

Whatever the theoretical arguments, CRINE, in conjunction with the technical developments described in Chapter 6, is clearly a great success. It is currently fashionable for companies to announce the staggering savings that they have made through using some new technical advance or some new system (be it alliance, partnership, changed internal management, or whatever.) In 1995, for instance, Phillips brought in the Judy and Joanne developments under budget in "virtually every area" – saving £35 million on platform procurement costs, as a result of "our procurement strategy, change control, contracting philosophy and control of materials"; £18 million on part of the drilling programme thanks to "teamwork" with the drilling contractor, and so on. In all the company estimates that it will save up to 15% on the budget of £767 million.[4] BP's alliance on the Cyrus field, involving Rockwater, Brown and Root and Barmac, expects to bring the Cyrus field on stream in July 1996 as a satellite of the Andrew platform, six months earlier than the target date, and £130 million under budget.[5] The number of such examples could be multiplied almost indefinitely from the pages of industry journals.

Annual reports and company press releases make similar claims, perhaps with shareholders in mind. LASMO's defence document against the Enterprise hostile takeover bid in 1994, for instance, claimed that LASMO's unit operating costs had been reduced by 18% between 1991 and 1993, with a further 17% reduction expected in 1994; staff numbers in the UK were cut by 20% in 1993 and administrative costs fell by 21% in 1993 and a further 20% in 1994.

One question which leaps to mind is why, if savings of this enormous level are suddenly so common, they were not made before. Perhaps the most obvious answer is Dr Gaisford's: the industry had grown accustomed to thinking that it had to be a high-cost producer, but the fall in the oil price in the late 1980s, once it became clear that it was a long-term fall, forced the companies to think again if they wish to survive economically.

The savings and cost reductions being achieved at the moment should certainly not be taken for granted. It does not have to be that way. The giant Hibernia project, to exploit a field offshore Newfoundland in Canada, is reported to have overrun its cost estimates by over C$1 billion. Delays in constructing the concrete platform at the centre of the project are likely to cause the operators to miss the 1996 weather window for towing out, with the result that production will not begin until 1997. Hibernia is quite a large field – it may contain as much as 800 million barrels – but in mid-1994 its operating company, Hibernia Management and Development Company, was reported to be C$1 billion in the red.[6]

It also seems clear that the high level of savings currently being achieved (or claimed to shareholders) cannot be achieved for ever. Consultants Arthur Andersen[7] argue that, although operating costs in the UK sector will fall by 15% to 30% in the period up to the year

2000, this fall is mainly made up by economy measures relating to the older platforms. If expenditures are projected over a 30-year period, Andersen see a total fall in capex of some 9% per barrel of oil equivalent, and in opex of around 6.5%.

There are some costs which, it is reasonable to assume, will increase. For instance, although the number of people employed in the industry is decreasing rapidly, the skill levels of those who remain is increasing, and it is likely that the costs per person employed will rise. Costs of safety training will rise; and as technology becomes more complex it will require higher skill levels to maintain. Similarly, the need for expensive environmental protection will reduce as new technology is used (for instance, an unmanned platform generates much less waste) but tighter legislation and increased ability to detect impurities may raise the costs of what remains. Abandonment costs will become a reality for the first time.

It has also been suggested that capex is being reduced in ways which will actually increase opex. Larry Famer, the president of Brown and Root Energy Services, said in late 1993 that operators worldwide would spend about $300 billion in opex over the next 10 years, as opposed to only $280 billion in capex. (Traditionally, he said, capex has been higher than opex.) In the North Sea, opex costs would rise due to the increasing age of facilities and the need to extend the life of reservoirs, as well as abandonment costs.[8] As described in Chapter 6, one result of recent technical developments is indeed to reduce capex (e.g. by chartering floating production facilities rather than building platforms, and by tying back to existing facilities such as pipelines owned by other operators rather than building new ones) at the expense of increased opex (e.g. the on-going costs of chartering the FPSO and of transport through the pipeline). To this extent some of the cost reductions currently being claimed may be misleading.

However, changing from capex to opex may be no bad thing. Opex costs in many cases are variable costs rather than fixed costs; they can be adjusted to match the price of oil and the profitability of an operation. What may happen is a shift in the ownership of resources, the capital requirements, and the degree of risk taken, from one part of the industry to another. If an operator charters an FPSO, it no longer has to make a major capital investment in a fixed platform. Instead, the investment is made by the FPSO owner, and becomes a quite speculative one: unless it can negotiate a lengthy charter agreement, it faces the possibility that the vessel may not attract sufficient work to pay for itself. The FPSO owner, like the drilling rig owner, has no control over the oil companies' decision on how and when to produce or drill. At the other end of the spectrum, if the FPSO market becomes tight, the operator may have to pay a very high charter rate if it wishes to continue production.

HOW ARE RELATIONSHIPS WITHIN THE INDUSTRY CHANGING?

> *The Third Driver: the Development of Relationships within the Industry*
> At best, there will be streamlined and synergistic relationships between operators, suppliers and other contractors and subcontractors. They will be well managed and driven by agreed common goals. In an unfavourable future, these relationships will be combative and inefficient: each company will seek its own interest, whatever it may claim to be doing, and overall management will be divisive and unable to control the process properly.

The "core competencies" approach, and the pressure to reduce costs, has led the operators to realise the extent to which they are competing in several business areas which they now regard as "non-core" areas. In the traditional pattern, the operator was responsible for selecting and managing every contractor who worked on a project, and the operator would also tend to specify the exact technology required. Now, rather than act simply as an employer of specialist firms (for instance of a drilling company) operators have tried to develop concepts of "partnership" in which the expertise of the specialist is used to the full.

This implies changes to the working relationships between the operator and the suppliers. It is intended to provide a better flow of ideas, and also to turn each project into a partnership where those involved share common goals and pool their resources to achieve them. In the new operator/contractor relationship, outside contractors perform a wide range of functions which were traditionally undertaken by the operators. Managing the construction of large projects and tying together the expertise required is the most obvious area. However, contractors also manage the operation of platforms; drilling contractors undertake and manage the drilling of wells. Some companies have outsourced such critical functions as accounting. As operators' head office staffs shrink, they are relying more and more on outsiders, not only for specialised engineering knowledge (such as how to design a crane) but also for fundamental management skills.

The most immediate issue, and the most controversial, is that the new style of agreement usually has economic implications for the participants. For instance, the older style of contract between a drill rig owner and an operating company is based purely on days worked. The operator gives detailed specifications for the well to be drilled, and charters the drill rig and crew from the owner on a day-rate basis until such time as the well is complete. A day longer on station means an extra day's charter cost for the operator. Under the newer agreements, such as Shell's "Well Engineering in the Nineties", the drill rig owner may be contracted on a performance related basis. Various complex formulae have been worked out to do this. The advantages are that it is now in the rig owner's interest to drill as quickly and effectively as possible, and it can use all its available skills to do so. The disadvantages for the rig owner are that it is now sharing some of the risk: if for reasons beyond its control (such as bad weather, or unexpectedly difficult geology) the well takes longer than planned, it may make a loss.

As well as a performance risk, the contractor may also take a commercial risk, if its return is partly related to the commercial success of the project. This is taking the industry into new waters. Contractors and supply and service companies have been forced to take longer-term perspectives on work they undertake. Larger contracting groups offer an ability to spread the financial risk which they are often now called on to share. It is perhaps only a matter of time before some of these "lead contracting" firms become partners in their own right in licences to operate.

Many supplying companies' advertising literature now describes how they have "grown internal expertise", or allied with external companies, to offer "one-stop shops" and to produce joint bids for major work. Good examples are Baker Hughes-INTEQ, Halliburton or the Dresser group, which offer a range of services from drilling technology and fluids to project management. Construction groups, such as AMEC Process and Energy, John Brown Trafalgar Oil and Gas or McDermott Offshore, offer similar spreads of capabilities in construction, production and operation.

Behind the new reward structures, however, it is not always easy to be sure what exactly is being changed. "Partnership" and "alliance" have become buzzwords, sometimes used with little thought of what they mean in practice. There are several different types of "partnership" or "alliance" systems. Some are more radical than others – and it is important to look behind the marketing rhetoric. In some cases, the arrangement is little more than a long-term pricing contract: the operator picks a supplier and negotiates a cheaper rate in return for a long-term commitment to buy. This may lead to savings and be heralded as an alliance, but lacks any element of cooperation other than on price.

In a more developed type of alliance relationship, the supplier and purchaser agree to work together to improve standards and design. Clearly, a long-term purchasing commitment is needed to make this viable, but the essence of the agreement is that it goes beyond price. Signs of such a relationship are the exchange of personnel, and certainly of ideas. Traditional bargaining over prices may be replaced by jointly agreed goals and standards to which both parties work. The resulting technology may become joint property. Rewards for both partners may be based on results rather than a fixed scale.

A more elaborate relationship still involves a customer (typically an operator) with several suppliers, working jointly on particular problems. Provided the suppliers are complementary and not competitive, they can jointly achieve a degree of synergy that might have been impossible under the old one-to-one customer–supplier relationship. An example might be drilling a well. Typically the well was designed, the drilling company chosen, and the specialised fluids bought, in that order. It might be more effective to take advantage during the well design stage of a specialised company's knowledge of downhole chemicals. Going beyond drilling, the alliance may also consider the well's production. The way a well is drilled, and the way it is produced, have traditionally been separate sets of decisions taken by different people. In fact, the way the well is drilled can have a considerable impact on the way that it later produces. A multi-disciplinary group is more likely to debate these issues than a series of experts reporting one by one to a customer.

The structure which has acquired the most notoriety is the most complex of all: that in which the operator hands over much of the management of an operation to a "lead contractor" who manages some or all of the other contractors. The advantages for the operator are simplicity: if the system works, it has only one set of bills to pay and one relationship to manage. The "lead contractor" is often a "one-stop shop" company offering a wide range of services, perhaps through a merger of smaller specialist companies. The "lead contractor" offers not only the synergy of cooperation between various disciplines, but also the management skills to achieve the customer's objectives.[9] It is important to remember that this sort of partnership is in fact offering two advantages: management skills, and synergy between the companies involved.

To manage a major activity, the "lead contractor" needs to demonstrate project management capabilities on a scale which carries credibility with the large operators. It is sometimes suggested that some of the "alliances" put together by suppliers are marketing devices with little real integration behind them: groups of companies jumping on the bandwagon, developing a new logo, but actually retaining their separate company cultures.

The offer of management skills is attractive to the purchaser because it enables it to cut out a layer of management from its own staff, and thus reduce its own costs. (Of course, the purchaser pays for the management time and skills provided by the lead contractor, but the assumption is that competition ensures that these are efficiently provided at a cost-effective price.) Secondly, it avoids duplication: it minimises the extent to which different management teams are watching each other. The offer of synergy is attractive because it suggests that a technically better product will be delivered, making full use of the expertise of the specialist suppliers.

The jury is still out on "partnership" approaches. Cynics among supplying and service companies may believe that they are just a device for the oil companies to push prices down and shift some of the risk to others. Optimists are more enthusiastic and believe that partnerships may lead to much greater efficiency. The Deputy Managing Director of AGIP UK, Jim Stretch, said in 1994 that "it would be unfortunate if we go down the path of partnering and it takes us nowhere. It could actually eliminate competition. Outsourcing contracts for the life of the field, how can you do that and be sure to remain competitive?"[10]

Mr Stretch's point is a good one. There are relatively few companies able to perform the "lead contractor" role. For a range of reasons, this is not a perfect market: it is difficult to compare one proposal with another, for instance. Long-term contracts are inevitable, but once they are signed, how can proper performance be guaranteed?

One key may be the extent to which proper management controls can be developed in these relationships. (A second may be the movement of critical personnel – see below.) In any large project it is difficult to keep track of what is going on. In a complex structure, depending on the relationship between several different company cultures and objectives, inefficiency and mistakes may be more difficult to identify and management controls even more crucial – but also more difficult to develop and apply. Simply looking at the overall result (e.g. in terms of savings against budgets) may not be enough. The opportunities for cost savings and efficiency increases are considerable at the moment – for instance the many technological developments outlined in Chapter 6. It may be that savings made for technical reasons will disguise, for a time, the real contribution of alliance structures. Whether these contributions are negative or positive may have a considerable impact on the future of the industry.

Negative effects might include disputes and even litigation; in such cases the flexibility of the partnership structure becomes a problem rather than an asset. If proper joint management controls are not developed, then either the project will be inefficient, or all parties will have to reintroduce their own supervisory layer. On the other hand, positive effects of these partnerships will be considerable: reducing costs, harnessing the talents of a limited number of skilled people in the best way, and minimising the time spent in intra-company squabbles.

As contracts become larger, layers of oil company management are relocated to "lead contractor" companies and the low oil price intensifies the struggle between suppliers, the importance of winning business is also growing. One large contract may be a matter of life or death to many supplier companies.

Dishonesty has not been seen as a major problem in the UK industry – certainly not in comparison with some other areas of the world where exploration and production have a more "Wild West" feel. However, the E&P Forum, an industry body, recently estimated that the activities of illegal information brokers cost the UK industry some £35 million per year.[11] Operating like intelligence agencies, these brokers obtain inside information on contracts and bids from company employees, or exert improper influence to twist the contracting process in favour of their "clients". More than 50 such brokers are believed to be operating in Europe alone. The estimated cost to the industry presumably represents the difference between a fair bid and a dishonest one: the Forum also estimates that illegal brokering can add 3–5% to the cost of a project. This would suggest that such brokering is thought to apply to around £1 billion of contracts per year, although no actual estimate appears to have been published of how much of the industry is affected. UKOOA estimated that the upstream oil and gas industry invested over £10 billion in 1992.[12] If so, illegal information brokering might be a factor in roughly 10% of all investment expenditures. The advantages of inside information about a rival bid, or the customer's evaluation criteria, are difficult to quantify; but anything which unfairly skews the selection procedure will inevitably diminish efficiency (and have adverse effects on morale).

It is possibly easier for the major operators to control such illegal activities, since they have sophisticated and experienced security departments. If they delegate many of the contracting arrangements to others, and if increased personnel movement loosens traditional company loyalty and allows insiders with knowledge to move more readily, it is to be hoped that newer entrants to the market will build up the capability to police illegal practices. Once again, this seems to be a case for good management controls. Making the tendering and selection process transparent and objective can do a great deal to reduce the impact of "inside" information, or even of favouritism (for whatever reason) inside the purchasing company.

WHAT WILL HAPPEN TO THE OPERATING COMPANIES THEMSELVES?

> *The Fourth Driver: the Industry's Own Internal Organisation*
> This includes the trend towards fewer people, with more skills, and the ability of organisations to learn. In a favourable future, the industry will adapt its own structure to meet the business environment and challenges it has to face. At worst, the industry will fail to adapt. In a high oil price environment, it may succumb to the temptations of "gold plating" – too many people engaged in non-essential activities. Or it may become overbureaucratic: perhaps in response to external pressures or in an attempt to maintain perceived quality, it will concentrate on bureaucratic "paper" controls
>
> ———— *continued* ————

> *continued*
> at the expense of real safety, quality and environmental effectiveness.

Cost cutting inevitably leads to concern about job losses. It has been estimated that over 500 000 jobs were lost worldwide in the oil industry between 1984 and 1994.[13] The early 1990s saw round after round of cuts in many oil companies operating in UK waters. The cuts were often accompanied by "rightsizing", "re-engineering", or "restructuring" programmes. Outside the major operators, the effects were less public but equally felt: some companies disappeared altogether, and others laid off workers.

Shell Expro's Production Director, Brian Ward, said in 1995, "The number of people offshore is one of the key costs of North Sea operations. People have to be transported by helicopter and expensively accommodated on platforms or floating 'hotels'. Yet we did little to control the number going offshore. People were sent offshore with little prior planning and coordination of their work. We have been able greatly to reduce the number of people offshore – saving costs, and improving communications and safety. Inevitably, people have lost their jobs in this process."[14]

The lesson of the expensive 1970s and 1980s has not been lost on others. In designing its Dunbar platform, which came on stream in December 1994, Total decided to go for as small a crew as possible. The platform was designed to be operable by the minimum number of people, with lifting aids and improved access built in. A semi-submersible tender was used to provide mud and cement facilities during drilling phases, in order to reduce the size of the permanent platform topsides and jacket. The crew were sent on over 100 separate courses, costing over £250 000, to make them fully multiskilled.[15] This emphasis on flexibility and formal training is notable throughout the industry. Offshore workers are given regular training courses in such subjects as safety and survival, as well as in their own jobs.

To some extent, claims that there has been a dramatic fall in numbers of people employed in the industry, are misleading, like the much-publicised job cuts by the larger operators. It is true that the larger companies are shedding staff; but in many cases these staff are then indirectly re-employed on the same or similar projects. Contracting firms, or smaller independent oil companies, may take on the running of platforms. A variety of functions may be outsourced, ranging from security and office services to keeping the company accounts and handling payments. The number of people employed in 1994 in the upstream oil and gas industries, onshore and off, was estimated by the Department of Trade and Industry at 300 000.

But it would be wrong to say that most personnel have merely been moved off one payroll on to another. There has been a real reduction in the numbers employed. Some of this is due to technological developments which reduce the need for offshore workers: not normally manned platforms, for instance, or the ability to make do with fewer structures.

Middle managers are also at risk. Large companies no longer wish to retain specialised expertise in house when outsourcing is seen as cheaper; although the daily rate of using a consultant may be higher, the consultant is taken on for a specific piece of work and no more, and does not accrue pension rights or rights to future employment.

In response to lower oil prices, some larger companies have cut down their activities, and some inefficient or unlucky smaller companies have ceased to exist, or at least have withdrawn from operations in the UK. The emphasis on new types of relationships between operators and contractors conceals a rapidly changing industry. The operator is still legally responsible for the field; but it may no longer directly control all activity on the field, or directly manage each subcontractual relationship. It may no longer specify custom-made equipment, preferring to buy "off the peg". All this makes room, in theory, for substantial cuts in middle management and specialised staff.

Large operating companies have traditionally been long-term employers. The Shell or British Gas employee, even 10 years ago, believed that he or she had a "safe" job. Each of the big companies had its own *esprit de corps* and its own way of doing things, its separateness nurtured by a strong sense of corporate identity and rivalry with its counterparts. The position has now changed, for all the operating companies and for most of those that serve them. Jobs are under pressure, and there are few employees or managers at any level who can afford to be completely unworried about their own futures.

Paul Gallaher of the AEEU said in late 1994 that the industry's drive to cut costs had led to thousands of lost jobs, corner cutting, a downward pressure on pay, and slumping morale.[16] The fall in morale has perhaps been most keenly felt in the larger companies, where the workforce had come to believe that their jobs were guaranteed for life. Workers in smaller companies were perhaps more hardened to economic realities. There are several anecdotal reports of such demoralisation in the larger companies. Immediately after the Shell Brent Spar débâcle (see Chapter 4) when the oil major was forced into a humiliating change of plan by the pressure group Greenpeace, a group of workers on Shell's Brent Charlie installation are reported to have sent a very public contribution to Greenpeace.[17] (Shell responded by sending warning notes to its employees that no donations should be made using the company name.) According to BP's in-house magazine *Shield*, the round of cuts in BP during the early 1990s saw head office staff reduced from 2000 to 350. The cutting and restructuring process became known within BP as BOHICA, or "bend over, here it comes again".[18]

The large companies had extensive in-house expertise and large bureaucracies. Some of them have been criticised for a "not invented here" mentality: the view that any development that was suggested by outsiders should be regarded with suspicion, while the in-house method was assumed to be superior exactly because it was "our own". One of the major management challenges facing the companies is to ensure that they remain "learning organisations", open to the outside world and able to identify the best wherever it comes from. They must also be flexible to work in whatever way seems to be most cost-effective.

It is here, however, that organisational history and a cost-cutting present may cause problems. To return to the illustration of CRINE at work, made earlier in this chapter, Vic Tuft's "cloud of uncertainty" includes oil company staff, who see their jobs disappearing at a rapid rate. Individuals have a strong incentive to argue that functions should not be outsourced, and to develop reasons, such as safety, quality, worldwide consistency or whatever, to support their beliefs. We quoted above the example of a CRINE case study in which the certification procedures for component parts of high-pressure items such as valves and manifolds were streamlined. Describing this case, the CRINE secretariat noted that there were "areas of resistance" in the client's purchase, quality assurance and certification departments – the ones whose jobs might be put at risk.

Managers have to understand this lobby. They have to persuade employees that CRINE, in Tuft's words, "is not about cost reduction at the expense of jobs, but extracting unnecessary costs which will help to create more opportunities, and so sustain employment levels at a higher level than would otherwise be possible".[19] If successful, managers also have to cope with the structural problems of re-employing the in-house experts.

There is a further management problem. There are only so many skilled people available. Competition for them may well increase. Large external suppliers are, in many cases, bidding for relatively few, high-value contracts. The Aberdeen office of a contractor, if it wins a large bid, will need to take on many more staff to cope. (If it loses the bid, even the existing staff who prepared the bid may be at risk.) Already, with smaller operating companies, it is not uncommon to see a project managed entirely by contractors, with the operator itself specifying the overall objectives. The actual expertise is highly mobile. Individuals may now work for a succession of companies: sometimes directly contracted to an operator, sometimes indirectly.

One key question is the ownership of the new skills and knowledge. There is no doubt that corporations can learn, and that their skills can improve over time. A fascinating study for the Gas Research Institute of Chicago compared the effectiveness of different operators drilling for gas.[20] By comparing the ROP (rate of penetration) of successive wells drilled by operators in the same area, it was able to demonstrate that there were considerable differences between operators in drilling effectiveness. Over time, the study showed that some operators became much faster whilst others stayed the same, suggesting that some companies could learn more effectively. As well as learning curves, however, some operators were also found to demonstrate "forgetting" curves: ROP slowed down when similar projects were taken on a few years later, or when key staff left and took the critical skills with them.

It is difficult to predict whether North Sea operators will show learning or forgetting curves in future. If the operators choose to allow the contractors to acquire more of the essential expertise, but do not allow them to maintain a sufficiently steady workload to retain the staff who embody that expertise, then it is likely that short-term cost savings will lead to a longer-term reduction in efficiency. Learning is likely to be thinly spread around the industry, largely in the heads of individuals. It seems clear that efficiency comes from a mix of skilled individual personnel with company cultures which recognise, understand and encourage the use of their skills, and that a degree of stability is essential to ensure this mix.

If the operators decide to retain a degree of expertise "in house" then they will have to accept the overheads of paying critical staff even when they are not fully engaged in defined projects, and the rigidities of a larger base of "permanent" employees. In other words, they may find that there is a limit below which they cannot "downsize".

Training may be another problem. When large operators expect to undertake large amounts of work in house, they have an incentive to train their staff to high levels. (It is noticeable at the moment how many specialists and managers in smaller operators were trained in one of the larger companies, particularly BP, and then moved out.) If more and more work is subcontracted out to contractors, or taken on by small niche operators, the industry may come to rely on these smaller companies to undertake most of the training required to maintain the overall UK skill base. But if the niche operators', or the contractors', business loads fluctuate widely, if they are forced to cut overheads such as training costs in order to remain competitive, and if as a result they live by snapping up a diminishing number of existing skilled individuals, who will provide training for the next generation?

WHAT ARE AN OIL COMPANY'S "CORE COMPETENCIES"?

The key issue for operators may be the definition of their "core competence". What skills do they retain in house? How do they define their competitive advantage over other operators? Is it that they drive a harder bargain with their subcontractors, becoming more and more an investment company with some specialist management skills – or do they retain certain abilities which will not be contracted out, and where they can point to their own excellence?

Areas of core competency might include:

- the ability to find oil and gas, and to spot opportunities to make a profit out of it. This is perhaps one of the easiest to measure. Estimates of finding and development costs (FDC) show considerable differences: over a recent three-year period, Phillips were the most efficient, with costs just over $3 per barrel. Mobil and BP were close behind. At the other end of the scale, with costs of $12 to $14, came British Gas and Enterprise. The three-year period is probably enough to average out the effects of good or bad luck.[21]

- the ability to raise capital. However, this involves different skills depending on the type of company. "Majors" may raise capital largely from their downstream operations; independents may be far more dependent on cultivating the City and on financial engineering. It is also not a specific oil industry skill.
- the ability to design and bring into being systems to exploit a field once it has been found. This may be no more than the ability to choose the best outsourced system and get it at a competitive price, or it may involve a strong in-house component.
- the ability to operate or manage oil and gas exploration and production projects once they are in being. Once again, this may be no more than the ability to choose good partners or contractors to undertake the operation. Esso, for instance, relies on its partner Shell to manage the majority of its operations in UK waters.
- the ability to process oil and gas and then to transport it between the field and the intermediate customer (e.g. the refinery). Again, this may be minimal: an FPSO would perform the processing and offtake functions, and some fields already have contracts in which the crude is sold effectively from the well-head to a customer – e.g. the Emerald field crude was sold on a long-term contract to Neste, which collected it from a floating storage unit in the field, the converted tanker *Ailsa Craig*.
- the ability to sell oil and gas at profitable prices, and to develop commercially successful contracts, hedges, etc. There are trading houses, the so-called Wall Street refiners, who do this very successfully without any direct involvement in exploration or production of hydrocarbons.
- the ability to balance the risks of a range of projects in order to deliver and manage a package of activities which is attractive to shareholders. Not all oil companies are good at this: but those which are not tend to run into trouble. The recent history of the independent sector provides several examples.

A sceptic might argue that all of these activities (except possibly the last) are already contracted out, to a greater or lesser degree, by some companies; or at least, that there are organisations which undertake to provide such services. Even overall management functions are not sacrosanct: oil and gas companies, like others, sometimes call in management consultants to advise them on strategy and its implementation. In practice each operator defines its core competencies in slightly different ways and the supplier market responds to fill the gaps. If there is one overall core competency, it is the ability to bring together all the different skills, control and manage them, and package the result in a way that makes money for the company and its partners and investors.

It is interesting to read oil company annual reports in this light: not all companies appear to have focused clearly on how they add value, not all say why they have apparently decided that some areas are "core" and others are not.

In the author's view, there is probably no reason why "oil companies" should undertake much current mature North Sea activity, even though traditionally they have had a near monopoly. (Some non-oil companies have been junior partners in licences – for instance ICI). If government permitted it, a producing field, for example, could be sold off as a package, with its management expertise, staff and contractor relationships intact. (There would, of course, be a problem about the ultimate liability for abandonment, but this could remain with the current operators.) Operators already consider contracting out the management of mature fields, and once this happens, the actual ownership seems increasingly academic.

Several companies, British and foreign, have made a speciality of buying diverse industrial companies, with proven processes, and then managing them at "arm's length". Especially if the oil price remains low, and if oil companies are short of capital, this sort of asset disposal may become attractive. For a buyer, the advantages would include a reasonably safe cashflow (provided the asset is carefully selected) and the ready availability of expertise, currently at very competitive prices, to manage it. For the seller, the advantages would include cash with which to return to its "core competencies" – and these might be to look for or develop new assets, rather than manage the old ones.

REFERENCES

1. Speaking at the Offshore 94 "Design and Development of Economic Hydrocarbon Production" conference held by the Institute of Marine Engineers/Royal Institution of Naval Architects; quoted in *Euroil*, April 1994, p 45. Amerada Hess estimate that production from their Hudson field will cost $8.63 per barrel, including an allowance for abandonment costs; so it can be done.

2. A G Kemp and B MacDonald, Cost savings and activity levels in the UKCS, *Energy Policy*, 1995, vol 23 no 1, p 71.
3. Article by Professor Kemp in *World Oil*, June 1994, p 37.
4. See *Offshore*, August 1995, p 49.
5. *Offshore*, July 1995, p 10.
6. *Euroil*, August 1994, p 110.
7. Quoted in *Petroleum Review*, April 1994, p 158.
8. Quoted in *Offshore*, February 1994, p 23.
9. These four types of alliance are described in an article by D C Coolidge in *Oil and Gas Journal*, 25 September 1995, p 58.
10. Quoted in *Euroil*, April 1994, p 36.
11. See D Knott, Shining a light on contract crooks, in *Oil and Gas Journal*, 10 April 1995, p 31.
12. *Petroleum Review*, April 1994, p 174.
13. *Oil and Gas Journal*, September 5 1994, p 41.
14. *Shell UK Review*, March 1995, p 25.
15. *Euroil*, January 1995, p 32.
16. Quoted in *Euroil*, December 1994, p 4.
17. Reported in September 1995, see *Petroleum Review*, October 1995, p 442.
18. A Sampson, Hats off to Company Man, *The Shield Magazine*, 3rd quarter 1995, p 31.
19. *Crinewatch*, published by the CRINE Secretariat, August 1995, p 2.
20. The study was conducted by Oil and Gas Consultants International and reported in *Oil and Gas Journal*, 25 September 1995, p 68.
21. Figures quoted in a talk at the Institute of Petroleum by Dr Richard Hubbard of Monument Oil and Gas in October 1995.

8
Market Forces

This chapter is about the drivers which come from the commercial environment. The most obvious are the demand for oil and gas, and the prices that consumers are willing to pay for them. Oil and gas are not standardised commodities, and there will always be price and demand differentials between different types of product. Changes in fashion or technology may make North Sea oil or gas more or less attractive over time. The market in oil and gas is not a perfect one: transport and infrastructure costs and developments can fundamentally change the value of a particular project. Even the nature of the markets themselves can have an impact. Last, strategic and political factors may alter the balance of trade, making indigenous supplies more valuable at times of international tension.

The President of Total, talking of the gas market in particular, summed up the two-edged nature of the markets:

> the rapid emergence of new market forces is very noticeable. This can be seen partly through the emergence of new entrants (the former utility companies such as British Gas, Enron and Nova, to name but a few, as well as the other non-OECD companies from, for example, the Former Soviet Union or China). It is also due to the increasing presence of the so-called market makers and the mutual funds trading on oil and gas paper. The market-place is crowded by all sorts of new competitors, resulting in an increased complexity and sophistication.[1]

THE OIL AND GAS PRICE

The most obvious commercial factor, and the one most often discussed, is the price. The earlier chapters of this book have tried to demonstrate that price (especially in gas contracts) is not everything; nevertheless, the overall price levels are a major influence on industry activity.

> *The Fifth Driver: the Oil and Gas Price*
> This includes the external factors that will influence it, such as world demand and events in other producing areas. In a favourable future, we assumed high prices in 2010 of $20 per barrel/20 pence per therm (in 1995 money terms). These would be stable price levels, with no recent or anticipated fluctuations of any size. At worst, we assumed a low oil price of $10 per barrel/8 pence per therm, or a price which might fluctuate from day to day but was not be expected to rise above this level in the medium or longer term.

THE INTERNATIONAL CONTEXT

The North Sea is only one of many oil- and gas-producing areas. In 1993 UK waters produced 3.2% of the total oil produced in the world, and 3.0% of the total gas.[2] This was roughly as much oil as Kuwait, Canada or Nigeria, as much gas as Turkmenistan or Indonesia. It was one-quarter as much oil as the USA or Saudi Arabia, or one-eighth as much gas as the USA.

The two giant oil-producing countries are Saudi Arabia (13.4%) and the USA (12.7%). Close behind is the Russian Federation (12.9%). (Figures for the former Soviet Union (FSU) are subject to considerable fluctuation and will remain so as those countries sort out their economies and their energy industries. Nevertheless, they have the potential to be even larger producers and consumers of energy.) There were 23 countries which each produced 1% or more of the world's total oil production in 1993: they included Egypt, China, Malaysia, Argentina and Algeria. 29.9% of the world's total was produced in the Middle East: "OECD Europe", the category which includes the UK and Norway, produced 7.7%, less than Africa (10.5%) or "Asia and Australasia" (10.4%). Oil consumption figures are, of course, another story. "OECD Europe" consumed 20.7% of the world total and the USA 25.2%. The Middle East, on the other hand, consumed only 5.6%, Africa a mere 3.2%.

The figures for gas are similar. The giant producers are the USA (25.3%) and the Russian Federation (27.5%). The Middle East is not a major producer. (However, the pattern of remaining reserves is rather different. The USA's reserves are almost exhausted: 7.4 trillion cubic metres (tcm) or 8.8 years production, compared with 44.9 tcm in the Middle East and 57.1 tcm or about 70 years in the former Soviet Union.) The 20 areas each producing more than 1% of the world total include Australasia, Indonesia, Venezuela and Mexico. Over one-third of the world's total production currently comes from the former Soviet Union, and just under one-third from the USA and Canada.

Gas is very expensive to transport. Either a pipeline must be built, or the gas must be liquefied at considerable cost before being transported as liquefied natural gas (LNG) in specially built tanker vessels. Because the cost of building a pipeline across continents is prohibitive, gas tends to be used in the region where it is found, while oil is traded and moved internationally and sometimes over long distances. In gas, therefore, regional consumption patterns are closer to production. The USA produces 25.3% and consumes 29.3%. The FSU produces 33.9% and consumes 29.8%. "OECD Europe" produces 9.9% and consumes 14.4% – the extra coming largely from the FSU. Africa produces 3.5% and consumes 2.0%. "Asia and Australasia" produces 9.0% and consumes 8.8%. (Within these regions, of course, there are smaller imbalances: for instance, Japan produces one twenty-fifth of the gas it consumes, while Indonesia produces five times more than it consumes.)

The UK oil and gas industry is just one among many. As a nation, the UK is lucky that it produces more oil (100.1 million tonnes in 1993) than it needs (84.1 million tonnes in 1993) – and nearly as much gas (production 56.7 million tonnes of oil equivalent; consumption 60.5). The wealth of OPEC countries such as Kuwait or Abu Dhabi is not due to the absolute size of their oil production (97.3 and 95.4 million tonnes in 1993 respectively – less than the UK) but to the fact that this is so much greater than their consumption, giving them surpluses to sell or withhold.

It is not easy to see whether the oil-producing countries will ever regain the short-term ability to control oil prices that they possessed in the 1970s and early 1980s. There are many new sources of both oil and gas, making it more difficult for any group of countries to monopolise production. The relative importance of the Middle

Eastern producers has declined slowly over the last 15 years: 33.5% in 1979 compared with 29.9% in 1993. The same is true for other regions: the USSR/FSU is down from 18% to 12.4%, although there are special reasons for this; the USA is down from 14.9% to 12.7%. Regions which have increased their share of the world's production include western Europe (3.5% to 7.7%), Latin America (8.7% to 12.9%) and "Asia and Australasia" (7.7% to 10.4%). Africa retains much the same share (10.5% to 10.2%) although there have been considerable swings within the region. At the same time, these figures do not represent a radical change in the world pattern of production: nearly one-third still comes from the Middle East, which still holds over 65% of the world's proved oil reserves. ("OECD Europe" holds 1.7%, the USA 3.1%, the FSU 5.6, Mexico and Venezuela 11% between them.)

New political forces make it less likely that OPEC will regain the degree of political cohesion required to act together to influence prices. Thanks to massive capital inflows from the sale of oil and gas, the Gulf Cooperation Council (GCC) states, Saudi Arabia, Kuwait, Bahrain, Qatar, Oman and the UAE, have enjoyed what one writer recently described as a "twenty year holiday from politics and economics".[3] From 1972 to 1981, the GCC's annual earnings from exports rose from under $10 billion to over $163 billion. After the oil price fell in 1986 these earnings also fell, to around $45 billion – but by this time the GCC countries had become used to their wealth and their expenditure on imports had risen. Saudi Arabia's government deficit is now 15% of GDP. Foreign assets built up between 1972 and 1985 have been drawn down and debts incurred. King Fahd announced 20% government expenditure cuts in 1994.

The former Soviet Union (FSU) has two different sets of problems. After the collapse of the USSR and the old-style communist leadership, the Russian Federation is experimenting with new political and economic systems and is not yet certain which it wants to adopt. The newly independent countries, in addition to defining their political and economic future, are striving to define their national identities and to achieve satisfactory relationships with Russia. As a result, oil and gas legislation and tax structures are unclear.

The second set of problems are more technical. The USSR's oil and gas industries were not developed efficiently. Production targets were achieved by draining a limited number of giant and supergiant oil fields, with the result that the reservoirs are producing below their maximum efficiency. Equipment is old. Oil companies are obliged to take on burdens which would be unthinkable in the West; in many areas the company provides regional infrastructure, social services, housing and shops for its employees. The oil and gas fields are a long way from potential markets or from warm-water ports, and pipeline access is often technically or politically difficult.

Yet the reserves of both oil and gas in the FSU are enormous, and the area has the undoubted potential to become a much greater world force in both commodities than it already is. How and when are difficult to foresee.

The USA continues to be both a major producer of oil and gas and the world's largest single consumer of oil (and joint largest consumer of gas with the Russian Federation). It is unlikely to continue to be a large producer for long: at the end of 1993 its proven reserves were 9.9 times annual oil production and 8.8 times gas production. Even allowing for further finds, in what must be one of the most actively explored countries in the world, US production must soon be expected to fall off drastically. The same cannot necessarily be said of US consumption: life-style expectations, and the demands of industry, do not go down without a struggle. There may be room for some savings from improved energy efficiency, although gross energy consumption per capita in the USA has fallen by very little over the period 1974 to 1992. (In 1974 the average inhabitant consumed, directly or indirectly, the energy equivalent of 7.9 tonnes of oil; in 1992, 7.7 tonnes. This compares with a world average consumption per person in 1992 of 1.5 tonnes, or a European Union average of 3.5 tonnes.[4]) US energy imports may reach 11 million barrels per day within a few years – five times UK production.

When it comes to the cost of incremental oil or gas supplies, the Middle Eastern producers have once again a distinct advantage. In many cases their reserves are under land. The costs of finding and developing new fields, or of improving the efficiency of old fields, may be less than $2 per barrel. Exploration and production costs in the North Sea, even for an extremely efficient company trimming its costs to the bone, cannot match this. (Some FSU reserves are also accessible from land, but in several cases they pose particular problems because they are in extremely cold areas or under permafrost.)

Oil (and to a lesser extent gas) are commodities which appear to disobey the laws of economics. The cheapest oil (such as that from the onshore giant Middle Eastern fields) is often held in the ground, while some of the more expensive (such as oil from the North Sea, Alaska or

Siberia) is produced and sold. Major Middle Eastern producers, with capacity to earn more than they need, have acted to keep the oil price high – thus in effect removing their own competitive advantage and allowing relatively high-cost producers a share of the market. It has been argued that Saudi Arabia's best way to increase its oil and gas revenues in the long term might be to overproduce today, and let the price collapse: first to stimulate demand for oil, and secondly to deter investment in high-cost fields elsewhere. This would drive out of business many other producers, including most of the UK offshore industry, which would be unable to attract investment for any new fields or field developments at an oil price of (say) $5 per barrel.[5] After a few years, the low-cost producers would have maintained their product's market share and achieved a monopoly of production. (This would be a curious reversal, turning the clock back to the "exploitative" production and pricing patterns of the 1950s and 1960s, but this time imposed by the producers and not the consumers.) However, Saudi Arabia and others would have to take the massive short-term cuts in revenue which would occur during the first few years of such a policy; this suggests that it is not a practical possibility.

There remains the spectre of political or economic upheaval: unrest or war affecting the FSU or the Middle East. In both areas, problems are clearly visible: nationalism and thwarted expectations in the FSU, and in the Middle East the triangular struggle between conservative undemocratic regimes, modernising liberal tendencies, and fundamentalism (sometimes accompanied by the ambitions of regimes in the area). The Iraqi invasion of Kuwait, and the international response, was to some extent misleading. The oil price did not alter greatly. But Kuwait is a small producer (less than the UK) and there was a quick and effective international response with the clear and achievable objective of restoring a legitimate government. With international help, the serious damage to oil fields and installations caused by the war was quickly repaired, and Kuwait's production in 1993 was higher than at any time since 1979. None of this is likely to apply if there is internal upheaval in Saudi Arabia or the Russian Federation.

HOW IS THE OIL PRICE SET?

The way in which the oil price is fixed has changed considerably over the last few years, even though superficially it appears similar. Crude prices are still determined by reference to so-called marker crudes: Saudi Light, Dubai Fateh, West Texas Intermediate (WTI), Brent Blend, and others. One change is that Brent is now used to price over 65% of the world's crude oil, although only some 700 000 barrels of it are produced per day, as compared with a worldwide production of around 65 million barrels.[6] (British newspapers referring to "the oil price" usually mean the Brent price. In the 1970s, the Saudi or Arabian Light price was the focus of UK attention.)

Since all crudes differ in their chemical components and their value to refiners, markers provide a convenient method of quoting prices. A price differential between any particular crude and a marker such as Brent Blend is determined. This will depend partly on its composition but also on such factors as the distance it has to be transported, and a negotiated element. Once the price differential has been accepted, the price of the other crude can be assumed to fluctuate with Brent, remaining a few cents higher or lower as necessary. Forward prices can then be quoted as "Brent plus $1" or whatever, using the marker as an index.

Plate 1 Loading North Sea crude; the tanker *Stavanger Boss* at BP's Hound Point terminal in Scotland in 1994 (Photograph by courtesy of British Petroleum)

However, the terms on which oil has been traded have changed over the last 20 years. Long-term contracts at fixed, or partly fixed, prices accounted for a large volume of trade in and before the 1970s; the spot price was in effect the price for small marginal supplies outside these contracts. The two oil price increases in the 1970s and 1980s broke up many long-term fixed contracts: as the price soared, desperate buyers were prepared to offer almost any figure. Sellers became very reluctant to enter into, or to keep, longer-term supply contracts. The spot market took over as a pricing mechanism: by 1982 more than half of internationally traded oil was priced by reference to the spot markets, even if it was traded on a long-term contract.

The Brent market is in fact not a "spot" market but three limited futures markets. In the "dated Brent market", cargoes of crude are traded for delivery 15 days after the contract date. (This is to allow the purchaser time to send a tanker to Sullom Voe to collect the cargo.) It is mostly a physical market: that is, most participants intend physically to deliver, or take delivery of, the cargoes. This market generates the marker price. It exists alongside two parallel Brent markets which are more often traded without physical cover, and allow traders to hedge or speculate. Given the large size of the traded contracts, however, they are a limited vehicle for speculators. In 1988 the International Petroleum Exchange introduced a futures contract for 1000 barrel cargoes. The New York Mercantile Exchange (NYMEX) allows similar trading. These new markets made futures trading easier and more accessible.

The Iraqi invasion of Kuwait in August 1990 led to a dramatic increase in futures market trading. The number of IPE contracts traded in August 1990 jumped by 40% from the previous month; NYMEX contracts jumped by 35%. For a few days, the logic of the world oil market broke down: physical holders of large oil stocks were reluctant to sell them. Asian buyers, heavily dependent on Middle East crude, were desperately short; the price differential between Brent and Dubai rose to $6. (In late 1995 it was roughly $1.30.) Arbitrageurs were unable to smooth out the difference, because crudes priced from Dubai were not available for sale for about six weeks.[7]

"Paper" trading is, in effect, speculation on the price of oil: traders arrange to buy (or promise to sell) cargoes of oil at a specified price on a specified future date, in the hope of making a profit if the price rises (or falls) from the levels for which they have contracted. These transactions may be conducted as a means of "hedging" – that is, physical traders may wish to insure against future adverse price movements by "locking in" with a futures contract a favourable price which is available to them today. In this sense trading is a prudent, even cautious, activity engaged in by most oil companies as a means of making their physical oil trading more manageable. Whether the actual "spot" price goes up or down, each physical trader knows what they will receive (or pay) and can work out their cashflow accordingly.

However, many "speculative" traders are no more interested in physically taking delivery of the oil than average horse racing gamblers intend to take the winning horse home with them: the oil market is just another place to trade. They hope to make money by, in effect, betting that prices in a futures contract will change by a certain range. (Thus a trader who expects the price to rise more than the market expects will buy forward contracts and hope to resell them at the higher price.) The presence of speculators helps the hedgers by ensuring that the market is liquid, but it also divorces the markets to some extent from the realities of physical oil movements; as the Kuwait crisis showed, quoted prices may be meaningless.

Plate 2 BP's oil trading floor (Photograph by courtesy of British Petroleum)

The growth of paper trading may actually skew the way in which prices are formed. OPEC controlled prices in the 1970s and 1980s by controlling the amount of oil that was available to trade. But it has now been estimated that, whereas OPEC may control 40–50% of the physical oil traded, it only controls some 5–10% of trading (i.e. of contracts written, whether for physical delivery or not).[8] As a result prices are as likely to be influenced by speculative operations on the New York Mercantile Exchange (NYMEX) as physical availability in the Gulf. What this will mean in practice is difficult to say.

THE FUTURE OF THE OIL PRICE

Forecasting the future price of oil is a masochistic activity, though one regularly indulged in by some finance houses. While economic modelling may allow reasonable confidence about short-term forecasts, there is no sensible technique for predicting the price five years ahead, let alone fifteen. As recent history has shown, the rules of the market have changed too often and too abruptly, either as a result of war or geopolitical changes, or because of changed consumer patterns, or short-term effects, such as warm winters, or decisions by large countries to build up (or run down) strategic stockpiles.

Even the irregularities are not consistent. There are interesting parallels with gold. The gold price rocketed in the early 1970s after President Nixon suspended gold convertibility of the dollar. At one time it was believed that gold was the ultimate means of storing wealth in difficult times, and the markets seemed to prove this: as international tensions and money-market uncertainties increased, the price of gold went up as well. Then this effect appeared to cease: crises came and went, and gold stayed relatively steady.

Similarly, the oil price rocketed in the early 1970s, as the producing countries forced a change in the terms on which they sold to the consuming world. First, there was a fear of shortage. The Club of Rome report, in 1972, predicted that oil would run out. In fact, since then some 500 billion barrels of oil have been produced, and the world's recoverable oil reserves now are half as high again as they were in 1972. Secondly, oil was used for the first time in 1973 as a political weapon, creating expectations that the price of oil might rise again in times of international instability, especially trouble in the Middle East. But the US bombing of Libya in 1986, and the Iraqi invasion of Kuwait in 1990 followed by a short but bitter war to throw the Iraqi troops back, had no long-term effect. (Brent crude prices jumped to $40 but soon fell back; the annual average price for 1990 was only around $24, falling to $20 in 1991 and $17 in 1993.) The first use of oil by producing countries as a political weapon against consumers has – so far – been the last. It is as if oil, like gold, has been removed from the financial and political arena and become a mere commodity, like copper or pork bellies.

Some reasons for this may be advanced. As argued above, the Middle East producers' dominance is slightly reduced in the short term. In a world where some major consumers (such as the USA) produce significant fractions of what they consume, the price may be driven by competition for the marginal fraction that has to be imported. There are few exporting countries which can afford to limit their own oil or gas sales, and existing attempts to set up producer cartels (such as OPEC) are riven by political dissension; even those quotas and price formulae which are agreed are often ignored by members. In 1973 there was, effectively, a cartel of producers who all shared similar political views. It is difficult to see the world's major exporters agreeing on any foreign policy issue at the moment.

Professor Peter Odell has argued that regional groupings may form between oil producers and consumers. Three such major groupings might be North America and Northern South America (i.e. joining the oil of Venezuela and Mexico, and Canadian gas, to the markets of the USA and Canada); Europe, including energy from the European FSU and North Africa; and Asia/Australasia with the Far East FSU. Professor Odell suggests that one driver for such regional groupings might be "fear of aggressive polices by the oil-rich and low cost Gulf exporters".[9] To some extent these groups already exist as oil and gas trade patterns, and some rudimentary steps have been taken to put them on a more formal basis (e.g. the European Union's Energy Charter).

In geopolitical terms, it would seem that the consuming countries have every interest in encouraging diversity of sources of supply. On the other hand, the pattern of the world's known reserves points inexorably to Middle East domination of the oil market, and FSU domination of the gas market – one day.

The demand side is also uncertain. New alternatives for oil and gas are available, and the technology is available to use older alternatives (such as coal) more efficiently. More efficient use of oil and gas may limit our need for both. Yet there have been no radical changes in patterns of hydrocarbon consumption between the 1970s and the 1990s, despite two sudden oil price increases. (Gross world consumption of oil and gas as a percentage of total gross energy consumption was 64% in 1974 and 59% in 1992, according to figures published by the European Commission.[10] Total gross world energy consumption in 1992 was 143% of that in 1974.)

While over time alternative fuels may change the relative position of hydrocarbons, the time and cost taken to substitute one fuel for another means that in any short-term crisis most fuel users are effectively locked in. It seems clear in the short term that the demand for oil and gas will rise. According to the International Energy Agency, the world will consume over 75 million barrels of oil per day in 2000, compared with 68 million in 1994. The number of cars is rising steadily. Outside the developed world, demand from newer and growing markets such as China is still only partially felt.

In planning our scenarios, we made a high and a low price assumption. On the low assumption, the oil price would have fallen to $10 per barrel by 2010, in real terms. (Gas prices were about 8 pence per therm.) On the high assumption, oil prices reached $25 per barrel and gas 20 pence per therm. In each case, we assumed that these prices were relatively steady long-term trends, rather than the result of wild fluctuations frozen on 1 January 2010.

We did not consider more extreme cases. It is possible that the oil price might fall lower than $10 per barrel; but if so this would almost certainly reflect society's judgement that it no longer needed to use oil in the same quantity or for the same purpose. A price at this level would lead to massive cutbacks in production – only major onshore fields, with much of their development already capitalised, would be likely to be viable. Such changes in our consumption patterns are not unthinkable, though they are unlikely within 15 years; but they would certainly change the face of the world.

Equally, it is possible to imagine the oil price rising far higher than $25: serious political instability in the Middle East and FSU leading to production cutbacks, and possibly to field damage requiring extensive investment and repairs before oil could flow again, would probably have this effect. Such a change would undoubtedly enrich some speculators and impoverish many others. It would slow world growth and, on past form, cause inflation and skew capital flows. Oil companies might initially rejoice at windfall profits, but it would be difficult for them to use these effectively – the temptation to "gold plate" operations would be too great to resist. Past performance by some oil companies, including diversification and overexpansion, has demonstrated this problem. Governments would rapidly impose windfall profit taxes, but for industrial nations such as the UK it is questionable if these would compensate for the other economic effects of a major price rise.

THE GAS MARKET

> *The Sixth Driver: the Development of the Gas Market*
> This driver includes the growing importance of gas in the offshore mix, the European gas market, and gas pricing practices. In a favourable future, there would be an increased demand for gas as the UK and mainland European markets expand, and/or gas captures a greater share. At worst, there could be dumping of imported gas, continuing price collapse, and chaotic fluctuations in the supply/demand balance.

The development of the UK gas market is a study in itself, and outside the scope of this book. There are three strands woven through the pattern; none of them is yet fully resolved. First, the rise and decline of a monopoly purchaser (British Gas). Secondly, the growth of the UK market itself, as gas gradually came to replace other fuels (originally coal, now also oil). New technology (such as combined heat and power gas-fired electricity generation, or gas-powered vehicles) may further increase the market. The growth of market size has been made more complicated by the deregulation of the market. Competition is driving prices down, and short- rather than long-term contracts are becoming more common; but it is difficult to see whether this will significantly expand the overall demand for gas. From our point of view, it is even more difficult to see if it will make the market more attractive to offshore gas producers. Thirdly, there is the question of whether the UK market is to remain a separate one, or whether it is to become part of the mainland European market. From 1998, the Interconnector gas pipeline will permit massive gas transfers between the UK and the European grid; whether they will occur depends on the supply/demand balance on either side of the Channel.

At the time of writing, the gas price in the UK is very low. A survey in 1995 found that prices to consumers had fallen by 40% in the year. UK prices were the lowest out of 13 countries surveyed.[11] Some offshore projects, such as British Gas's Olympus field, have been delayed. British Gas itself is said to have entered into forward price contracts which commit it to pay for some £500 million of gas for which, because of growing competition, it now has no market. At the time of writing, it had taken its main counterparties (Shell and Esso) to court to argue the interpretation of gas purchase contracts from the Strathspey and Brent fields. The British government was urging all concerned to go for "sensible commercial renegotiation" of the contracts. This is a sensitive issue; if not well handled it may leave a legacy of distrust in the market which could hold back gas developments offshore for some time. It may also hasten the demise of complex long-term "take or pay" agreements, and increase reliance on the spot market for gas sales.

Demand in the UK alone is unlikely to require large new external gas sources before 2005. UK developed reserves are estimated at 1275 billion cubic metres (bcm) and undeveloped reserves at 640 bcm; production is expected to decline from 2000 onwards. However, continuing price falls, uncertainty about contractual or marketing arrangements, or simply the unpredictability of relying on a spot market for sales, may slow down investment and therefore future production.

The UK is now embarking on an extremely radical programme of gas market liberalisation. Although mooted in the 1980s, when British Gas was first nationalised, the liberalisation programme has required successive government interventions to force a gradual reduction of the British Gas supply and marketing monopoly, and to allow other companies to supply increasing numbers of industrial consumers in the UK. As from 1996, domestic consumers in south-west England will also be able to buy gas direct from other suppliers, who are already promising to undercut British Gas prices. Eventually every gas purchaser in the UK will have a choice of supplier.

Evidence from the US deregulation of the gas market suggests that such changes of the rules take a while to settle down.

For gas producers, market liberalisation is already having an effect. The East Brae field was reportedly the first example of a commitment to the development of a gas field without a field life depletion contract in place – in other words, of the producing companies assuming the marketing risks of a major deposit of gas on their own.

Meanwhile, the UK gas market is still a local market. There is already one pipeline connecting the Norwegian Frigg field to the UK, and about one-third of UK gas supplies were imported through it during 1977–87. (Such evidence as there is suggests that this "foreign" gas was bought at much higher prices than the "captive" gas from UK waters.[12]) The future of this line, and whether it can be used to transmit gas into or from pipeline systems on the Norwegian side and thus form a connecting link with Norway, is still under discussion between the two governments.

The Interconnector subsea pipeline is currently under construction between Bacton in the UK and Zeebrugge, with the purpose of linking the UK into the European grid. When completed in 1998 it will be capable of exporting 20 bcm or importing 10 bcm per year. (The directional difference in capacity is due to the compressor layout.) The likely profitability of this pipeline for its operators is still under debate. In order for it to transport gas in either direction, there has to be a sufficient demand, and sufficient price differential, to make UK gas plus transport costs marketable in NW Europe (or vice versa). Price and demand are difficult to forecast on either side of the English Channel.

A well-developed international gas market now exists in parts of continental Europe, based on a growing international and national pipeline grid. The EU and EFTA consumed 284 bcm in 1993, of which 239 bcm was locally produced. There were major flows within Europe, from the Netherlands to Germany and Belgium, and from Norway to Germany and France. In addition, gas is imported from outside Europe (mostly from the FSU to Germany, and from Algeria to Italy.) Norway is taking a lead in developing pipelines to supply Norwegian gas to France and to north-western Europe. (The European market is still limited: Spain and Portugal remain relatively isolated, for instance.)

The future supply and demand balance for mainland Europe has been examined in detail by several experts. Andrew Wright, of MAI Consultants, concludes that European gas demand in 2010 may be as high as 630 bcm or as low as 470 bcm, from a current level of 350 bcm for EU, EFTA and eastern and central Europe. European developed gas reserves were 6000 bcm in 1994: 10% in the UK, 32% in the Netherlands, and 46% in Norway. Current EU/EFTA production rates are 239 bcm. Approximately 38% of gas consumed is now imported. Production of natural gas within the EU is expected to decline after 2000, leading to requirements for new sources estimated at 70–130 bcm per year.[13]

There are several potential new sources: the most obvious is increased supply from the FSU. Reported inefficiencies and underproduction in the FSU suggest that, given the right political and economic conditions and the necessary infrastructural improvement, very large amounts of gas could be quite easily made available: this has been referred to as a massive "gas bubble".[14] (It has been estimated that the amounts of gas used in the FSU as pipeline turbine fuel, or lost in transmission and storage, are roughly equivalent to the overall level of UK gas production.) Other large gas fields are available in North Africa. It seems to be generally accepted that there will be no technical difficulty in supplying the additional gas which Europe requires; but it is less easy to predict the diplomatic and security implications of relying increasingly on the FSU and North Africa for supplies.

On one analysis the UK gas producers could find that when they finally break out of the limited UK market they enter another market which is equally oversupplied; a "gas bubble" in mainland Europe could flow back down the Interconnector and further reduce prices in the UK. On the other hand, UK producers may find that a healthy market expansion in Europe will lead to firmer prices and a more rational market in the UK – whether or not they actually sell gas outside the UK.

For these reasons, it is difficult to say which way gas will flow through the Interconnector. One argument is that its main effect will be to transmit prices between the UK and the European gas grid. Once it is possible for gas to be supplied from one end to the other for a given transport cost, then prices at the two ends ought not to diverge by much more than that transport cost. Gas does not actually have to flow for this effect to occur.

In our scenarios, we assumed a gas price in 2010 ranging from a high estimate of 20 pence per therm, and a low estimate of 8 pence per therm. However, the real assumption was that, in the favourable scenarios, the market for natural gas (whether in the UK alone or in the whole of Europe) would develop in a way which would encourage UK offshore gas developments. This implies a degree of stability which will make investment possible. Price instability may lead to a vicious circle. (Prices fall – investment falls – supplies fall – prices rise, etc.) The low estimate assumes either that such instability occurs, or that stable low prices continue, perhaps as a result of oversupply and less demand in the UK and Europe than optimists expect.

RISK PROFILES AND THE AVAILABILITY OF MONEY

> *The Seventh Driver: the Availability of Money and the Development of Risk Management Techniques*
> In a favourable future, capital will continue to be available for projects in UK waters with a competitive rate of return. More risk management tools will be developed to help borrowers and lenders structure their risk portfolios. At worst, a shortage of capital and a perception of high risks might curtail all but the most obviously profitable developments. Other geographical areas or industry sectors will appear more attractive to lenders. Risk management tools will lose credibility.

Serge Tchuruk, the President of Total, has said: "More and more, managing an oil company means trying to master the art of hedging against a volatile environment. But this is easier said than done."[15]

The upstream oil and gas industries are all about risk. Exploration expenditure is a gamble that oil will be found where the geologists think it will. Development expenditure is a further gamble. What makes the risk more important is the enormous size of the stakes. Thanks to modern risk management techniques, companies now have a greater ability to decide on the level of risk which they, or their shareholders or bankers, want, and to fix that level using market instruments.

Sometimes, of course, the judgements made by management are just plain wrong. The oil and gas industries are no more immune to managerial egos and wishful thinking, or to bad luck, than any other. In an industry which has had its share of "colourful" personalities and spectacular good luck, there is perhaps more tolerance than there should be of the strong character who has been lucky in the past.

Oil and gas companies take a range of commercial or financial risks. These include:

- political risk: the risk that the host government may change the rules. At best such changes may be modifications to the tax regime, such as the removal of PRT in the UK; at worst, they may include

nationalisation or the enforced rewriting of contracts to reduce the company's rights or profits.
- exploration risk: the risk of finding "dry holes".
- reservoir risk: the risk that a reservoir, having been identified and shown to hold oil or gas, may turn out to be smaller or less productive than originally thought. This was, for instance, the case with the Emerald field in the North Sea: thought in 1989 to contain 37 million barrels of recoverable oil, it was found in 1991 that the oil contained far more water than expected. Recoverable reserve estimates were downgraded to 15 million barrels.
- market risks: the risk that oil or gas prices, or demand, may fall below the levels at which the project makes economic sense. This is, for instance, the reason for British Gas's delay in developing the Olympus gas field: the company already has commitments to buy more gas than it needs.
- other commercial risks: the risks of changing interest rates, rising prices, default by partners, delays or problems in construction, major accidents, and so on. Some of these can be covered by insurance, but by no means all. The costs to Occidental of the Piper Alpha disaster, for example, appear to have been covered by insurance. Cost over-runs during modification of the Emerald Producer floating production facility led to significant difficulties and litigation for all concerned. British Gas "take or pay" contracts for gas which it did not take – see above – are reported to have been worth at least £500 million for one quarter in 1995. The company is now arguing the interpretation of future contracts.

Commercial risk is perhaps the most complicated. Enormous profits may be made from the discovery and exploitation of a major oil (or gas) field – provided that all the risks are properly managed. It is also possible to make large losses. The *Oil and Gas Journal*'s list of the top 300 publicly traded oil and gas companies with production in the USA (the "OGJ 300") shows the effect of a lucky find on company growth. In 1994, for instance, the fastest growing company on the list increased its income by 1298%. Seven of the twenty fastest growing companies increased their incomes by more than 100%. All were small. (Revenues for the 300 as a whole grew by only 2.1%, whilst profits fell by 11.9%.) The OGJ 300 does not single out the fastest falling companies, but its tables show some dramatic drops in revenue.

Any development may be broken down into four phases. The first phase is pure cash outflow. Exploration is relatively cheap, but of course brings in no revenue. The largest slice of investment is the development of a field, especially when this requires the construction of a platform and related infrastructure. There are various techniques, technical and analytical, for reducing the risk at each stage. However, it is not until the second phase of the development, when oil (or gas) begins to flow, that the investor receives a return. This, typically, is the period of highest payback. The "easy" part of the prospect – typically the largest part of the reservoir, under natural pressure, with no complications – is now producing at full volume and the development costs have been covered.

The third phase shows production or revenue falling off, as the reservoir ages, or as smaller pockets of oil are developed which cost more to recover, or as water begins to be produced. Some additional investment may be needed now, to enhance recovery, but if well judged it still brings in good returns. In the fourth phase, the effort involved becomes greater than the returns: the cost of continuing to operate the facility is growing (ageing equipment needs to be repaired or replaced) and further investment in enhanced recovery does not bring in enough revenue to cover costs. During the third or fourth phases of the curve, the company should either abandon the prospect or sell its interest out.

In real life the boundaries between the four phases are not so clear; it requires a mixture of commercial and technical judgement to decide where one is on the curve, and much depends on the assessment of risks. There are also a range of strategies: for instance, a company may specialise in buying assets in the third and fourth phases of the curve, and in finding ways to maintain production at reasonable cost. An exploration company may specialise in developing prospects through the first phase of the curve, bringing in other investors to help with the massive costs and retaining only a small share in the rewards from the second phase. Some companies argue that they have a low cost base which allows them to make a commercial success of projects which others cannot handle. At the opposite end of the scale, only a few companies have the resources and borrowing power to handle the really large developments. Each strategy has its own set of risks.

Commercial risks have always been limited or spread in some way. Some risk-management options are as old as the oil industry, and consist largely of diversification: oil companies generally do not have all their eggs in one basket. Political risk is spread by operating in more than one country. Exploration and reservoir risk are spread by taking partners in one's own operations, and taking shares in other companies' wells. Companies also try to balance their portfolios to include both high-risk but high-return exploration projects, and low-risk, low-return shares in known, developed, revenue-generating fields.

Newer risk management options are apt to be much more complex. Oil price hedging is a relatively straightforward procedure (though getting it right is not). There are also far more complicated instruments devised by banks to help their oil industry clients to manage their risk profile.

As an example, an independent oil exploration and production company may have borrowed large sums of money to develop a particular asset. This loan is likely to be at a floating rate of interest. In the worst case, if the oil price falls as the interest rate rises, the company may be forced into default. It can insure against this by "integrated" or "composite" hedges, linking interest rates and oil or gas prices. For instance, the rate of interest on a loan may be held at an agreed low level for periods in which Brent crude remains below an agreed price. When the crude price rises, the interest rate also rises. These extremely complex hedging structures allow the company to decide on the level of risk it wants to take and (for a price) to offload its remaining risk on to a "counterparty" which may be its bankers, or may be the markets on which its bank is in turn hedging.

Complex uses of the financial markets have achieved notoriety in recent years. Some companies, in and outside the oil industry, appear to have over-reached themselves, getting into complicated arrangements which they themselves did not fully understand. Banks have an interest in persuading their customers to use the more sophisticated financial instruments. (The banks charge management fees for devising and implementing them; they are also a means of attracting customers and locking them in.) At the same time, new control systems are being developed which enable senior management to keep track of the company's overall exposures to the markets. It seems essential that, when complex hedging and other instruments are used, management is prepared to monitor them carefully, and that they can be shown to be part of a rational and cost-effective strategy.

There is no such thing as a "safe" oil company with a guaranteed profit, except perhaps those national oil companies which in effect collect rents on a national asset. One attraction of the oil and gas business is the risk/reward balance that it offers the investor and the participant. This cannot be designed out altogether, partly because of the inherently speculative nature of an industry dependent on finding hidden resources, and also (one suspects) because all concerned like it the way it is.

Increased involvement of the financial markets may sometimes tend to encourage uncertainty. This has been seen most recently in the currency markets, for instance in the speculations by George Soros and others against sterling, which forced Britain to leave the European Exchange Rate Mechanism. It has also affected commodity markets: for instance the Hunt brothers' attempt, with Saudi backing, to dominate the world's silver markets between 1973 and 1980.[16] If nothing else, the complexity of new instruments, and of the Treasury operations of international oil companies, has increased the complexity of management functions and made new demands on senior personnel.

According to Serge Tchuruk of Total, "If anything, the volatility of the oil markets, both short term and medium term, is due to increase even beyond the already high level already experienced."[17]

THE AVAILABILITY OF CAPITAL

Finding capital to develop and operate offshore has traditionally been done in two ways. The larger integrated oil companies have massive downstream sales networks, producing a steady cashflow from sales of gasoline and other products at filling stations and elsewhere. Such companies can and do finance developments out of their own resources. Shell, for instance, claims over $70 billion of capital employed, and at a return of 10–12% on this sum would easily earn enough to finance its major upstream projects. (These were listed by its group treasurer in a 1994 paper as requiring a total expenditure of $16.6 billion between 1993 and 2000, $1.7 billion of which was for the Nelson project in UK waters; the remaining projects were elsewhere in the world.)[18]

Major companies – even Shell – do borrow. For instance, although Shell's net debt to equity ratio was estimated to be 3.3% in 1995, Texaco's was estimated at 44.7%, Chevron's at 35.9%, and Exxon's at 33.7%. Very large companies with solid asset bases, however, have a range of options for raising capital. They can make traditional corporate borrowings, secured against the company's asset base or cashflow. If a project undertaken by a large company fails, then typically it is a small part of the company's operations. Individual failures have less effect on the company's enormous bottom line, as long as the overall percentage of success holds up, (although the enormous capital exposures of major projects are becoming so large that even the majors seek to offset some of the risks. Shell and Exxon, for instance, are 50/50 partners in most of their activities in UK waters).

The smaller independents, who typically have few assets other than exploration licences and some producing platforms, run a much higher risk if they borrow heavily against their existing assets. For a smaller company, developing one field might be a very large part of the company's operations. If the field does not produce as well as expected, the result might be to bring down the whole company; the bankers could have recourse against the company's other assets to meet loan repayments. The only option, for many small companies, was to trade out of a field or sell their interests to a larger company before the major expenditure on development had to be faced. This meant that it was difficult for an exploration company to take full advantage of a successful find. (Ranger Oil, for instance, had considerable difficulty in the 1970s in raising its share of the capital required to bring the Ninian field on stream, to enable the company to keep its investment and profit from the eventual revenues when the field went into production. Banks were not then willing to lend on the strength of oil that might be in the ground; and Ranger was a small company with little else to put up as collateral.[19])

The development of limited recourse (or non-recourse) financing has enabled funds to be raised with far less risk to the operator: in effect, the lenders share the risk. In limited recourse financing, the loan is linked directly to the hydrocarbons and cashflows of a particular project, and secured against them. It is particularly important where field sizes are relatively large and expected production lives relatively long (which has often been the case in UK waters). It means first that a small company can obtain a loan at all, on the basis of its expected share in the field rather than its current assets. Secondly, it means that if the project fails, the bank only has limited recourse, say to the company's share of revenues from the project, and not to the other assets of the company. The company balance sheet looks healthier and the shareholders' assets are less at risk.

Costs of such financing were initially very high – one of the reasons for the majors often preferring to raise finance in simpler ways. However, costs have fallen over the years, with the banks prepared to take a greater share of the risks and becoming increasingly familiar with the problems involved. In the majority of cases they appear to have judged correctly. A few companies, such as Midland and Scottish Resources plc, have had to come to voluntary arrangements with lenders as a result of disappointing performance from fields that were financed by project loans.

The financing instruments have become much more complex, and both lenders and borrowers vie to develop more ingenious devices for transferring or minimising risk. The net result is that capital is available for oil and gas exploration: for instance, syndicated loans to the oil and gas sector worldwide have risen from $25 billion in 1986 to $60 billion in 1994.[20] One estimate, however, is that the worldwide future investment needs of the petroleum sector will be up to $300 billion annually, compared with $200 billion annually over the last five years. About one-quarter to one-third of the estimated new investment is for upstream activities.[21]

Whatever the overall size of the pool, the UK industry will have to compete for a share of the available capital with the oil industries of other countries. The UK offshore industry has several advantages in the competition for capital. First, UK tax rates are low. Secondly, the UK is a politically safe country, unlikely to nationalise investments or to experience the horrors of civil war. Its current government appears keen to encourage production and (apart from the PRT changes) to ensure a stable tax and regulatory regime which allows companies to plan ahead. There is no reason to think that any alternative government would adopt a radically different approach. Thirdly, there is a great deal of infrastructure and knowledge; the UK offshore fields are well studied and represent a relatively safe bet. New provinces (such as, for instance, Namibia) are largely unknown; they might hold supergiant fields, but they might absorb large amounts of capital for little or no reward.

Against this, the larger oil companies need to find larger fields – the "cash cows" of the future. If it becomes the received wisdom that there are no more such opportunities in UK waters (except possibly west of Shetland) then developments here may become the province of smaller niche players, who may be less able to attract capital in a competitive environment.

186 Waves of Fortune

The market for energy financing is now an international one. It is more innovative, but also more complex. It can be expected to provide more flexible finance, on better terms and with flexible allocation of risk, to companies which understand how to use it. As with any other sort of high technology, however, the possibility of a disaster in the hands of the inexperienced, unwary or simply greedy is likely to be higher.

HOW LARGE ARE THE REMAINING UK RESERVES OF OIL AND GAS?

> *The Eighth Driver: the Extent of Britain's Untapped Oil and Gas Reserves*
> Where will any new reserves be found, and how large will they be? In a favourable future, new fields of reasonable size will continue to be found in frontier areas, and smaller fields developed in the areas which have already been well explored. Licence acreage will be fully exploited. At worst, the new frontier areas will be disappointing. Companies holding licences will sit on smaller fields and prevent them being developed by others.

How much oil and gas remain under British waters? This is a critical question for the industry, and indeed for the country as a whole. As we have seen, in the early days of the industry most commentators underestimated the UK's resources.

In the mid-1990s, estimates are much more optimistic: it seems as though, given favourable conditions, the industry will be able to continue to produce hydrocarbons for at least the 15 years covered by the Scenario Planning Workshop. While production rates may fall off as the giant fields are exhausted and their place is taken by smaller and shorter-lived fields, the Scenario Planning Workshop assumed that rates would stay, in favourable conditions, at over 1.5 million barrels per day of oil, and the gas equivalent of 1 million barrels of oil, in 2010.

Official British government figures, given in the 1995 Department of Trade and Industry *Energy Report*, express recoverable oil and gas reserves remaining at 31 December 1994 using a range of estimates. Proven reserves are those which are already specifically discovered and fairly well surveyed. Remaining proven reserves of oil are some 575 million tonnes (about 5 years supply at the 1994 production rate of 126 million tonnes per year). Remaining proven reserves of gas are some 660 billion cubic metres (ten years' supply at the production rate for 1994 of 65 bcm per year).

The next layer of estimates are "probable" reserves: a further 920 million tonnes of oil, and 855 bcm of gas. The third category quoted are "possible" reserves. In these cases, there is specific evidence that the reserve may exist, but the size or viability is not proven. Possible oil reserves are a further 580 million tonnes; possible gas reserves a further 400 bcm. Lastly, undiscovered recoverable reserves have not been specifically identified, but are estimated based on statistical techniques. Undiscovered reserves are estimated within the ranges of 1160–4580 million tonnes of oil, and 430–1602 bcm of gas. It is therefore clear, given the DTI estimates, that there is likely to be enough oil and gas under UK waters to provide for all or most of the national need until 2010.

The Department of Trade and Industry also estimates where the remaining reserves are likely to lie. The two largest areas for oil are the north and central North Sea (estimated range from 400 to 1640 million tonnes) and west of Shetlands (680 to 1750 million tonnes). For gas, the largest estimated reserves lie in the southern basin, Irish Sea and Celtic Sea basins, at 230 to 690 bcm; next comes west of Shetland at 150 to 570 bcm.

It is generally thought that all the major fields in British waters have already been found. (The main exception to this judgement is the area west of the Shetlands.) The remaining areas, particularly in the North Sea, have been relatively extensively surveyed and drilled. This does not mean that all the prospects found have been developed, or even that all possible prospects have been found. Several "lucky" finds have been made in areas where the geology was thought to be well understood. The Morecambe Bay gas fields had been a British Gas preserve for many years; it took other companies to find large oil deposits in the area. The Alba field was found by accident, by a well going down to the nearby Britannia field.[22] The Nelson field was developed by Enterprise Oil after several other companies had looked at it, but no one had realised how large it is. Scratch almost any geologist and you will find a pet theory, or a fond belief in a formation that no one has looked at properly. There are many known or suspected plays which have not yet been developed, usually because other plays were more attractive.

The area west of Shetland, which is in much deeper water, has not yet been studied in such detail and there remains the possibility of

188 *Waves of Fortune*

Plate 3 Exploration west of Shetland: the "Ocean Guardian" drilling rig conducting an extended well test in BP's Foinaven field (Photograph by courtesy of British Petroleum)

large finds there. As techniques for drilling and operating in deeper water improve (see Chapter 6) these prospects become available for study. If more giant fields like Brent or Ninian exist in UK waters, this is where they are likely to be. Whether or not the DTI's statistical methods, and its estimates of the potential recoverable reserves in this area, are correct remains to be seen. The oil companies are certainly behaving as though they agree with the Department: 26 exploration blocks were awarded in early 1995, with one block having eight competing bids made for it. It is estimated that more than 1700 sq km of 3D seismic will be shot west of Shetland in 1995, with even more to come in 1996.

For most operators, however, the preferred route is to look at smaller prospects, which are now becoming economic as recovery costs fall. "Small" is a relative term: projects now under development or coming on stream include Britannia, with an estimated 2600 bcf of gas, and Captain, with an estimated 350 million barrels of oil. New technology and lower cost structures may enable stand-alone fields as small as 5 to 10 million barrels of oil (or the gas equivalent) to be viable, and even smaller pockets of oil or gas to be developed from existing infrastructure.

OFFSHORE INFRASTRUCTURE

> *The Ninth Driver: the Availability of Offshore Infrastructure*
> This includes both technical availability, and the commercial terms on which existing infrastructure will be available to new users. In a favourable future, the existing infrastructure would continue to serve in new capacities. There would be no technical failures, and companies would make infrastructure commercially available to each other, enabling the development of fields and projects which would not otherwise be viable. At worst, the infrastructure would be prematurely abandoned, or there would be serious failures; infrastructure owners would make it difficult or unattractive for outsiders to share. This would seriously limit the number of satellite fields, or small fields in the main North Sea areas, which could be developed.

Chapter 6 examines in some detail the techniques by which existing offshore infrastructure, such as pipelines and platforms, can be used to make new developments much cheaper. Deviated drilling from an existing platform, or tie-back from a subsea completion to an existing platform, can make the difference between viability or not for a small field.

The critical commercial question is the terms on which the owner of the existing infrastructure make it available to the new developer. Each new development may involve several existing ones: drilling from a platform, tying back to a pipeline, using compression or separation facilities on an intervening platform, reception and treatment at a shore terminal, and so on. In effect, the infrastructure owners have a stranglehold over the new development; they can make it possible, or kill it before birth. This is more than a commercial question, since it may have a considerable effect on the overall development of the UK's offshore resources over the next 15 years and beyond.

190 Waves of Fortune

As a result, the British government has now issued a draft code of practice for companies to use in negotiating access to infrastructure. Its key proposals include more transparent tariff structures, and "non-discriminatory negotiated access which will not favour any particular company".[23] It has been issued as a discussion document and hopefully will help the industry to arrive at an agreed position. Perhaps one factor in favour of such agreement is the network of investments in the fields: a company owning a share in one new development is also quite likely to own shares in an older development and its infrastructure, if not in the same field then in a similar situation elsewhere. There is endless room for bargaining, and the role of government is perhaps to ensure that no large operator is able to develop a stranglehold in any area.

REFERENCES

1. Serge Tchuruk, President Director General of Total, quoted in *Petroleum Review*, March 1995, pp 111–14.
2. Figures from *BP Statistical Review of World Energy*, June 1994. Oil and gas production and reserve statistics are produced in different ways by different countries, and are not all entirely reliable or directly comparable. Figures given by BP and the *Oil and Gas Journal* are generally accepted as the best that are publicly available.
3. V Zanoyan, After the Oil Boom, *Foreign Affairs*, November/December 1995, p 2.
4. Figures from European Commission (DG XVIII), *1993 – Annual Energy Review*.
5. *The Economist*, 18 November 1995, p 120.
6. F E Banks "Oil and Money" *Energy Policy*, 1994 vol 22 no 12, p 993.
7. Figures from *The Economist*, 12 January 1991, pp 84–5.
8. Edward N Krapels, President of Energy Security Analysis Inc, quoted in *Oil and Gas Journal*, 25 April 1994, p 31.
9. P Odell, Global energy market: future supplies potentials, *Energy Exploration and Exploitation*, 1994, vol 12 no 1, p 59.
10. *1993 Annual Energy Review*, published by European Commission (DGXVII), April 1994, p 26.
11. Report by National Utility Services, quoted in *Petroleum Review*, January 1996.
12. See J D Davis, *High Cost Oil and Gas Resources*, Croom Helm, 1981, p 147. Not only was the gas price reportedly "many times" higher, so was the load factor.
13. Estimate by Domenico Dispenza of Snam SpA; see *Oil and Gas Journal*, 13 March 1995, p 45.
14. See J Stern, *The Russian Natural Gas Bubble*, R11A, 1995.
15. Serge Tchuruk, President Director General of Total, quoted in *Petroleum Review*, March 1995, pp 111–14.
16. See S Fay, *The Great Silver Bubble*, Hodder and Stoughton, 1982.
17. Serge Tchuruk, President Director General of Total, quoted in *Petroleum Review*, March 1995, pp 111–14.

18. Figures from a paper given by Stephen Hodge, Group Treasurer of Shell International Petroleum Co Ltd, to an Institute of Petroleum Conference on "Financing the International Oil Industry" on 13 February 1995.
19. Private discussion with a Ranger director.
20. See A S Whyatt, Director of Hardy Oil and Gas, quoted in *Petroleum Review*, April 1995, p 154.
21. Estimate by ABN Amro Hoare Govett, as presented by John Martin, Director of Corporate Finance, to a conference at the Institute of Petroleum on 13 February 1995.
22. See *The Chevron Magazine*, Summer 1995, p 3.
23. Department of Trade and Industry press release, P/95/344, 22 May 1995.

9
External Influences

This chapter looks at a set of drivers that are for the most part beyond the control of oil and gas companies. Although the companies may have some ability to influence them, all but the last are essentially "political" factors: decisions about the relationship between the state and industry, about the value we place on our environment, our society, and our national industrial heritage and position. They include the actions of the British government and of the European Union, our role as consumers and members of the public, and the effects of the "environment".

THE POLICIES OF THE BRITISH GOVERNMENT

The Tenth Driver: the Actions of the UK Government
This driver includes tax rates and licensing policies. Here we assumed that the "best" option is for continuity of existing policies on tax and licensing, with any changes being gradual and with their impact on the industry well thought through. The worst option would be sudden changes, with unexpected effects, leading to uncertainty and reducing the ability of companies to plan ahead.

British government policies now influence most closely the taxation of companies, and the initial licensing and subsequent development of fields. Other areas, such as safety legislation, are driven increasingly by European Union policy.

Tax Policies

One of the most complex drivers, but also one with an immediate impact on profits, is the tax regime under which the oil companies operate. Heated arguments took place in the 1970s between the Labour government and the companies, as the former strove to define an effective tax structure and the latter to claim a profit level that would satisfy their requirements. Mutual political suspicion did not help.

The Conservative government was able to take higher taxes, with less objections by the companies, because the dramatic rises in the price of oil made everyone's calculations much easier until the mid-1980s. Conservative ideology favoured minimal intervention by government in industry and simpler taxation; though it is not until the mid-1990s, with oil and gas prices back down to where they were (relatively speaking) in the early 1970s, that special taxes on oil and gas production are being phased out, leaving Corporation Tax as the main instrument.

In the mid-1990s, the UK government takes less in tax from exploration and production by oil companies than does any other country except Ireland. This has been reported by several studies.[1] The state take from fields is estimated at 33% in the UK; this compares with 25% in Ireland and much higher figures in Norway, where "economic" fields pay 87%; "marginal" fields pay less.

Oil and gas industry representatives often argue for a "progressive" system, which allows rapid cost recovery but an increased state take as a field matures. This encourages new field development, while taking more from economically stronger operations. "Regressive" systems discourage new developments, taxing new or marginal fields more highly, often by offering ungenerous allowances for development expenditure. (Petroconsultants estimate that tax rates in Syria are around 86.8% for extremely profitable fields, and over 100% for only marginally profitable fields.) UK tax rates are estimated to be the same for any type of field: neither progressive nor regressive but neutral.

Arguably, British tax policy in recent years may have been too progressive. The March 1993 UK Budget announced the removal of

Petroleum Revenue Tax for new fields, and the abolition of exploration and appraisal relief against this tax (with some transitional arrangements to smooth the effects of the change). After these tax reliefs, structured to encourage drilling, were first introduced in 1983, the post-tax-rebate cost of exploration had been estimated at 17p in the £; this measure does appear to have encouraged exploration activity in UK waters. (Depending on your point of view, some would say it encouraged too much drilling.) The effect of the 1993 PRT change was to raise the post-tax-rebate cost of drilling from 17p to 67p in the £. Exploration activity fell as a result, by 27% in the year to March 1994. (It can be argued that it would have fallen anyway because of the low oil price and high drilling costs. Similar tax changes in Norway and the Netherlands also led to low exploration activity in these countries during the same period.[2])

There are two reasons any subsequent administration might wish to rethink tax policies on the offshore industry. The first might be if the oil and gas price rose (or fell) by a considerable amount. Then a windfall profits tax (or, if prices collapsed, some form of relief) would seem to be essential. A second motive might be a belief that, even at existing prices, the balance of taxation was wrong: that the state was not taking enough. In such a case the government might be wise to set targets for the investment levels it wished to attract to UK waters, and to monitor the effect of any tax changes (or proposed changes) on investments and statements of investment intentions. This would not be an easy task, since there would be a great deal of complex argument on both sides; but it might prevent a slowdown in activity which could take years to reverse.

In our scenarios, we assumed two possible outcomes. The favourable outcome was that UK tax rates and structures remained pretty much as they now are: stability was seen as almost as important as the absolute level of rates.

In the unfavourable outcome, UK tax rates were changed in a way which hastened the closure of economically marginal fields, and curtailed exploration activity. (In other words, that left only the most profitable fields still in operation.) Such policies would drastically alter the face of the North Sea. Overall production would fall. There would now be a significant multiplier effect: basically profitable fields would be killed off if the plug was pulled on infrastructure on which they relied. International oil companies would turn away from UK waters to other more attractive areas. The optimum development of resources would be skewed by "bottom line" factors. If the remaining fields were developed to maximise short-term returns from easy reservoirs, and if longer-term investment in marginal developments, enhanced recovery programmes or field improvements was discouraged, then a significant part of the oil or gas in the fields would become almost irrecoverable.

In practice, most governments do not behave as if they had an entirely free hand in setting tax structures and rates. The oil industry is very international. With the exception of some state oil companies, almost all exploration and production companies spread their portfolio across many countries. It is generally recognised that there is competition for oil company investment, and that a country which wishes to see its resources developed must offer a competitive return. For instance, in 1993 the Norwegian government warned of possible tax increases. The industry replied that Norway might be pricing itself out of the market. In 1994 the Norwegian Prime Minister, Mrs Bruntland, in what was described as a "major U-turn", announced that it would not increase taxes after all: "My government realises that the ability to obtain adequate profit levels affects the outlook and planning of oil companies."

A study by consultants Van Meurs suggests that governments, when establishing terms for oil exploration and production, are more influenced by regional neighbours than by global comparisons (though UK rates are clearly not on a similar level to Norwegian ones!). Another factor may also be the perceived nature of the reserves offered. If these are potentially large, companies may be prepared to pay more for the possibility of finding a large field on which the long-term economics are good. If the majority of developments are likely to be small (as now in the UK) tax becomes more critical. The way in which tax is spread over the life of the field is also critical: "front-end loading", or taking a higher tax rate in the first few years than in the final years, is common practice. It benefits those operators who find big fields, while discouraging the development of smaller resources. Discussing the Van Meurs study, the *Oil and Gas Journal* says, "regressivity and front end loading suggest ways that governments can improve terms as world-wide competition for exploration capital intensifies. They can counter the trend by improving terms for marginal prospects relative to others, and by not claiming value from production until companies have recovered costs."[3] In other words, the timing and allocation of the tax take, and the way it is divided up between different types of operation, are as critical as the absolute level or the percentage rate.

Governments have a powerful if somewhat blunt weapon, in the tax system, for influencing the way in which their resources are developed; they also have a means of encouraging or discouraging investment. If significant changes are made to the taxation of the offshore oil and gas industry, it is vital that the blunt weapon should be made as accurate as possible, and the effects of its use thought through carefully. This suggests a continuing need for close consultation between government and industry, so that each side understands the other's priorities. Our worst-case assumption was that the tax weapon might be used indiscriminately, possibly as a result of a dogmatic ideological approach, or because of too close government attention to short-term fiscal results.

The Importance of Consistency

Tax is something which can be altered almost overnight; there are no lead times or engineering problems. Governments have the right to act unilaterally, although there are in practice constraints on their freedom to manoeuvre. A reputation for consistent treatment of investors attracts investment; but such a reputation can quickly be lost by an ill-judged move.

It seemed as though the abolition of PRT in 1993 was such a move. It was heavily criticised by some operators, particularly independents with a high focus on exploration activities: Ranger Oil, for instance, said that it had received $33 million of PRT repayments in 1992 and had expected (but would not receive) a similar refund in 1993; as a result "it is expected that [Ranger's] North Sea exploration will be curtailed".[4] According to *Offshore* magazine, "the word Britain was virtually unspeakable for oil companies".[5]

The Deputy Managing Director of AGIP UK, Jim Stretch, described the implementation of the PRT change as "awful", but added: "it seems that at last the government recognises that there is nothing special about the profitability of the upstream industry which qualifies it for additional taxation".[6]

Larger operators, however, generally welcomed the change; large mature fields enjoyed a boost to their post-tax cashflow.[7] And as *Offshore* magazine went on to admit, "Then came a collective change of heart. The same companies suddenly decided in 1994 polls that Britain was one of the most attractive oil regimes in the world."

It seems that the UK government has in fact judged the mood of industry correctly. Exploratory drilling trends in the UK in the mid-1990s have included more appraisal wells, often drilled at low cost from existing structures to probe the extent of known fields or to find satellite structures. These are often cheaper (given an unskewed tax policy) than wildcats, and often more likely to succeed. New fields continue to be developed, and at the time of writing the demand for offshore drilling rigs is high.

There will always be a tension between short-term macroeconomic management decisions which governments are forced to take, and the long-term investment appraisal which oil companies carry out. When they are making their investment decisions, companies have complete freedom to come to UK waters or not, and the government needs to woo them if it wants them. Once the major investments have been made, the companies are in effect hostages, obliged to accept whatever return they can get. (It is not quite so simple in practice, since of course a development typically demands several tranches of investment at various times in its life, and later ones do not have to be made, or made in full.) The development of UK gas fields illustrates this. The early developments continued to produce despite disappointing returns; but once these low returns were established other fields which had been discovered, but for which development funds had not been committed, remained untouched for years.

British Government Licensing Policy

Licensing policy is a means by which the government controls the rate and the location of exploitation. Licensing rounds are only held at certain times, and only some blocks, pre-selected by government, are offered. This is, of course, a game of chance, since the government does not know what is under each of the blocks, any more than the oil companies do. It can only try to choose blocks (or "tranches" of blocks) to ensure that the apparently most favourable areas are not "cherry-picked" to the exclusion of others. At the moment, government policy is to use the system to encourage and extend development and production.

Once blocks have been licensed, an agreed exploration programme takes place. If economic prospects are found, development plans must be submitted and agreed. The government in effect has complete power, at any stage of a field's life, to curtail (but not increase) activities or production. It seems unlikely within the next 15 years that any British government would wish actively to prevent development by prohibiting it through the licensing system, or to limit exploration by cutting the number of blocks offered for new licences, or altering the form of licences to make them unacceptable.

The rate of exploitation of UK oil and gas reserves is not affected by European Union decisions. It is entirely within the national competence of the UK how much, or which areas, of its sea bed or territory it offers for licensing for oil and gas exploration, provided that it conducts its licensing process in the manner approved by European union policy (see below).

Issues which may resurface are the length of time for which licences are valid and the relinquishment clauses. These have varied in past licensing rounds. Allegations are made from time to time that some of the larger holders of acreage are sitting on blocks which they have no intention of developing. It is sometimes suggested that such blocks should be released to other applicants, who would be prepared to develop them. To some extent the nature of the licensing system has encouraged applications for large areas. The cost per block is relatively low. Typical relinquishment provisions may require the successful bidder to give up half of the acreage awarded after a period of (say) six years. This, it has been argued, encourages large bidders to bid for twice the acreage they want, knowing that they will surrender the least interesting half of it when called on to do so. Since the relinquishment is half the acreage, not half of each block, the licensee can more easily retain all the areas with potential and abandon the rest.[8] (But Canadian experience of relinquishment by block – i.e. half of each block, not half of the total acreage granted, had to be given up after a time – was that a patchwork quilt of holdings soon developed. This was difficult to operate and administer.)

Whether some acreage holders do "sit on" blocks is a question that the author is not competent to answer. There may be many good reasons for developing a company's portfolio of blocks at a given rate: not least that endless investment capital is not available. (Nor has there been endless demand, particularly for gas.) Development may have to await progress on infrastructure in a nearby field. It may be influenced by tax considerations: awaiting revenue from an existing field, against which new development expenditure can be offset.

Relinquishment terms have tightened, and licence periods shortened, over the years. Allegations of companies sitting on blocks became an issue under the Callaghan government in the UK during the late 1970s; but it seems as though, if the present Conservative government believes that blocks are being "sat on", it prefers to work by quiet persuasion to release them.

British Government Policies on Safety and the Environment

The British government, in common with many others, has taken an increased interest in safety and environmental issues over the last few years. However, these are areas where Europe is taking an increasing lead.

THE INFLUENCE OF EUROPE

The Eleventh Driver: the Influence of the European Union
We included European Union rulings on competition, integration, and directives on working hours and the Social Chapter. In a favourable situation, we saw that there would be little extension of these and similar measures. In an unfavourable world, we saw new requirements and rulings leading to increased costs, limiting managerial flexibility and imposing new bureaucratic requirements. We were conscious that much would depend on the industry's own reputation, but also on the ability of policy makers to consider fully the likely impact of policy measures and the economic implications they will have.

In the late twentieth century, the authority of the nation state has been weakened in several ways. One of the most controversial, in the UK, has been the ceding of a wide range of powers to the European Union (EU). The successive treaties binding the UK to the EU have imposed obligations to accept European Directives and to enshrine them in domestic legislation. These have affected most areas of British life – from the way in which butter is sold at supermarkets to major questions of industrial policy.

The European Union has had some impact on the UK offshore industry, although British governments have resisted EU involvement more in this politically contentious industry than in most others. The UK has by far the largest offshore industry of any of the EU twelve. (Norway, with a comparable industry, did not take the opportunity to join.) Relations between London and Brussels have been tense. As far back as 1975, the European Commission attempted to limit UK subsidies to Scottish companies, while at the same time offering to subsidise exploration in the Rockall area.[9] More recently, the UK government has resisted Commission attempts to treat the industry in the same way as its land-based counterparts, without recognising its special characteristics.

Some European Commission (EC) measures have been generally welcomed. The Hydrocarbons Licensing Directive of May 1994 sets common rules for EU countries when granting off- or onshore exploration and production licences. The main intention is to minimise discrimination on any national basis in issuing licences. Although 20 years ago British government policy was briefly to favour "national" companies such as BP, the UK sector has perhaps been the most open to international competition ever since. By contrast, several European countries have allowed virtual monopolies of national companies or consortia, and are now being forced to open these up; a process which may well benefit British operators and suppliers. In Denmark, for instance, all oil and gas was initially produced by the Danish Underground Consortium (DUC), a private consortium made up of the Danish company A P Møller with partners Shell and Texaco. Later, to attract more foreign capital and to regain a degree of state control, licences were opened up, but with mandatory participation by the state company DOPAS. The EC directive will place limits and constraints on this participation; this is a sensitive major economic issue for Denmark. Norway, although not a member of the EU, is affected by the reciprocity provisions of the Directive. Where a non-EU state does not allow EU companies to win its licences on a non-discriminatory basis, the Directive gives EU countries the power to refuse licences to companies 'effectively controlled by that state or its nationals'. This could vitally affect the Norwegian state oil company, Statoil. The EU took the unusual step of inviting Norway to the discussions on this Directive before it was adopted in early 1994, to reach an acceptable compromise. One of the most immediate losers from the Directive has been Italy: large areas of Italy (e.g. the Po Basin and onshore Sicily) have been effectively a monopoly for Italian companies.[10]

As a result of the Licensing Directive, the UK Department of Trade and Industry must publish details of licences for which application is invited, at least 90 days in advance in the EU Official Journal. It must adhere to specified criteria in judging the licences. In practice, these conditions are not greatly different from those used before the Directive came into force. They include the technical and financial capabilities of the applicants, the need for proper exploitation of the field, and the commercial terms. Countries have wide powers to control licences for strategic, economic or safety reasons; but the main change is that they may no longer favour national companies or attach procurement clauses requiring, for instance, that a certain percentage of the money spent on the field is spent locally.

Directive 93/38, of June 1993, for Coordinating the Procurement Procedures of Utilities Operating in the Water, Energy, [etc] Sectors, has had a mixed reception. It imposes similar conditions of transparency on all contracts over a certain size issued by "public utilities", which includes the holders of UK petroleum production licences. In other words, not only does government have certain obligations when awarding licences to companies; the companies, when they become licensees, also become subject to similar requirements in their purchasing procedures. (Details of contracts over a certain size must be published, and qualification, bidding and decision procedures are affected.)

For an industry which has probably not discriminated on national grounds, but has on balance placed its contracts where it saw good value, this new requirement is often seen as imposing an additional bureaucratic load. There are also some suggestions that it may interfere with the development of alliances and partnerships (see Chapter 7), since the loose and interactive grouping of an alliance would not allow rigid tendering and evaluation processes to be carried out.

The EU has taken a considerable interest in the liberalisation of European gas markets (see Chapter 8). The success or failure of this policy may have a considerable impact on the demand for gas and the means by and price at which it is sold. The Commission has pressed for third-party access to national gas networks. It has encouraged "unbundling" – the separation, in vertically integrated companies, of the functions of production, transmission and distribution. (The division of British Gas into separate companies for these functions, and the growing amount of gas supplied to UK consumers by other companies, are examples of these principles at work.) One effect of this liberalisation has been to open up the possibilities of foreign markets for UK gas supplies – assuming that the gas can be transported there. The EU programme of trans-border cooperation on energy networks (Regen and Interreg) is likely to increase the demand for natural gas by making supplies available in new areas: through main gas grids in Spain and Portugal, and a gas link between the UK and Ireland.

The Energy Charter, signed in 1991 by nearly all European countries including eastern European and former Soviet Union countries, as well as Australia, Canada, the USA and Japan, is intended to liberalise trade in energy, and to promote and protect investments. It is also intended to spread good environmental and safety practice and to disseminate research and know-how. Although such grand measures are necessarily vague and limited in their impact on actual transactions, the Charter is one of the first attempts to establish a regional framework for energy security, based on the recognition that both supplier and consumer countries benefit from stable trade. EU aid and programmes such as TACIS and THERMIE have been specifically targeted at the energy needs of eastern European countries.

The EU has also examined such questions as whether the proposed Interconnector gas pipeline between the UK and Belgium interferes with competition law, and it has considered the possibility of setting up EU crisis measures in case of any shortage of gas supplies in the future.

A Commission Green Paper on energy policy was published in early 1995, with the three main objectives of improving energy competitiveness, security of supply and environmental protection. Although these ideals are generally accepted by all, there is some disquiet in the energy industries generally about the ways in which the Commission may propose to achieve them. The Commission argue that if Europe does not have an energy policy, then it is reducing energy to a mere commodity, and one which will run the risk of being "simply the handmaiden and servant of other Community policies".[11] However, UK government spokespersons, such as the Deputy Energy Secretary at the UK Department of Trade and Industry, Charles Wardle, reply that the European Commission already has sufficient powers, and that there is an adequate EU energy policy which does not need to be extended. The British government view is that most developments in European energy affairs have been managed adequately without a "European" involvement. This is in effect not a legal debate about how the existing EU rules need to be altered, but a political debate about whether they should be altered – in other words, about the proper extent of European power.

On a national level there are, of course, major differences in interest. The Netherlands and Denmark produce more natural gas than they consume, but only the UK produces more oil than it consumes. Germany, for instance, with an annual consumption of 131.13 mtoe of oil and 56.80 mtoe of natural gas in 1992, produced only 3.55 mtoe of oil and 13.72 mtoe of natural gas. (In 1992 UK production was 95.91 mtoe of oil and 45.57 mtoe of gas, against consumption of 83.28 and 50.17 mtoe respectively.)[12] It is therefore in the interest of "have-not" states to tie UK production firmly in to the EU and in the interests of "have" states to resist attempts to limit their freedom of manoeuvre. States which rely heavily on non-EU energy sources have more incentive to support the Energy Charter process.

A considerable amount of lobbying also takes place on behalf of the energy industries. Certain proposed measures, such as a carbon tax, would (if ever adopted) have extensive implications, and it is not at all clear how the impact of these measures could or should be assessed. Placing a tax on fuel use, based on the amount of carbon use involved in individual fuels, would alter the relative commercial terms on which fuels are used. "Clean" fuels such as natural gas, which produce little carbon on burning, would attract less tax than others such as crude oils, and the heavier products made from them such as fuel oil. It has so far proved impractical for the EU to introduce a carbon tax directly, largely because of the probability that it would penalise European industry in international competition by raising energy prices. (In the author's view, it would provide a prime example of the bluntness of the tax weapon.)

Whatever the future of this most complex of arguments, the EU remains like a distant overhanging snow slope. The possibility of an avalanche is always there. The supranational power of the European Union is potentially enormous: such a major development as a carbon tax would completely change the face of Europe's energy industries in ways which cannot be predicted. In our "unfavourable" scenarios we assumed that the avalanche might have fallen in some way.

Until the avalanche is triggered, however, the effects of the EU are relatively small and are felt mostly in such areas as safety, the environment and research.

Safety is an area in which much national legislation is based on EU directives. These are increasingly being applied offshore. (See Chapter 4 for a discussion of current UK safety legislation which is based in part on the European Council Directive on (Extractive Industries) (Boreholes) of 1992.) Other EU proposals, however, have aroused more resistance, given the special character of work offshore. A good example is the EU Working Hours Directive, which limits the maximum hours that can be worked. If applied to offshore activity (against the wishes of the British government) it could be argued that shortening the time spent by individuals offshore would increase the overall frequency of travel by helicopter, which is statistically one of the most dangerous periods for an offshore worker. In this sense the Directive would be counter-productive in the North Sea; it would also put the operators' costs up to an entirely unrealistic extent.

The present British government's opposition to the "Social Chapter" is well known; the Labour party is, however, pledged to allow it to be adopted by the UK. If implemented in the North Sea, it would have a radical effect on the role of the workforce, including a requirement to have Works Councils. This change would, of course, affect all other British industries. Some industry figures have argued that it would reduce competitiveness[13] – but this is a debate which goes well beyond the offshore oil and gas industries.

In the area of research and development, the EU has funded several initiatives. Between 1975 and 1994, for instance, the Commission claims that it has offered financial support for 3003 technology projects in the fields of energy conservation, renewable energy sources, and clean coal and hydrocarbons technology: a total contribution of more than ECU1.8 billion. This has been made most recently through the THERMIE programme. Projects may either innovate or disseminate technology. Preference is given to those involving collaboration between unrelated organisations in more than one member state, or which disseminate experience to less developed regions of the EU. Financial support is up to 40% of the costs of "innovative" projects, and 35% of the costs of "dissemination" projects.

Projects supported in the oil or gas industries as part of the THERMIE programme include exploration (such as 3D seismic developments), production (such as composite construction technology, or submersible vehicle design) pipelines (such as leak detection methods) and environmental protection (such as reducing the effects of sea bed scouring around small structures at sea). Various projects have looked at platform decommissioning methods – e.g. systems for cutting up platforms at sea using explosives, abrasive jets of slurry, or cable sawing.

CONCERN FOR THE ENVIRONMENT

> *The Twelfth Driver: Environmental Issues*
> This includes the impact of any serious incidents that may occur, the development of social attitudes, and the impact of existing and new regulations. In a favourable future, the industry would continue to pay serious attention to its environmental responsibilities. A social consensus would develop that allowed all concerned to agree on the general level of these responsibilities. Industry would be seen to be pulling its weight. At worst, there would be division over what is and is not acceptable, making it impossible for industry to meet agreed standards. Either industry would become overcautious (placing too much emphasis on environmental issues) or some companies would cease to take these issues seriously and would do no more than they can get away with. A major environmental disaster might occur; public reactions would lead to tighter controls and more suspicion, which would limit investment and flexibility. New fuels, made more economically viable by environmental pressures, might take over from hydrocarbon fuels in some areas.

Different Patterns of Energy Consumption

One broad "environmental" question is the extent to which patterns of fuel use may change. This may be owing to economic causes (e.g. if a fuel suddenly and permanently becomes much cheaper) or to tax alterations. It may follow changes in the way in which society perceives fuels, for example the extent to which we decide to minimise the pollution caused by fuels. Oil was seen, some 30 years ago, as a cleaner alternative to coal. Now it is more likely to be gas, hydroelectric power or "alternative" sources which are welcomed as cleaner alternatives to oil.

Europe's final energy consumption in 1992 was, according to European Commission figures, 226.36 mtoe for industrial purposes, 246.39 mtoe for transport, and 309.98 mtoe for "domestic and tertiary" use.[14]

In the industrial and domestic/tertiary sectors, the use of oil is generally decreasing, partly being replaced by gas. Between 1974 and 1992, industrial use of oil fell from 39% of the total to 19%; gas rose from 22% to 35%. In the domestic and tertiary sectors, oil consumption fell during this period from 49% to 31%, while gas rose from 18% to 37%. Competition between fuels is increasing: gas is becoming more readily available, new fuels such as orimulsion are being introduced, and old fuels such as coal are being revived through new technologies. The fuel buyer has an increasing range of choices, and competition appears to be driving prices down, at least in the short term.

In the transport sector, there is still little choice. Many technically successful attempts have been made to replace oil-based fuels in at least some parts of the industry. (Possibly the only type of transport where hydrocarbon fuels are indispensable is commercial aviation. Despite Yorkshireman Sir George Cayley's heroic experiments with steam-powered aircraft in the nineteenth century, no one has succeeded in commercial powered subspace flight without using petroleum products such as gasoline or kerosene.) However, road vehicles powered by electricity have been used for specialised purposes for a long time, and many attempts have been made to make the technology more flexible. Bus and taxi fleets, and some private vehicles, run on LPG in the Netherlands. Ships have used nuclear fuel, or high-tech "sails", to minimise their use of oil-fired engines.

All these options are viable, but none has yet become a significant challenge to oil. European Commission figures show that in 1974 the EC countries used some 142.44 mtoe of fuels for transport purposes; 138.29 (97%) mtoe came from oil. In 1992, the total was 246.39 mtoe, of which 242.4 mtoe (98%) came from oil; in other words, oil's domination of the market has grown slightly.

Hydrocarbons are not the perfect fuel. When burnt, they leave residues, partly from the fuel itself and partly from additives contained in it. Gasoline, for instance, requires additives to improve its performance. Although the use of lead-based additives has been dramatically reduced in most European countries on health grounds, some alternative additives (such as oxygenates like MTBE) are still controversial. Diesel fuel cannot function without a small amount of sulphur, which acts as a lubricant; it also tends to emit particulates. Even LPG causes problems when used as a transport fuel. While producing less pollution, it is more difficult and dangerous to handle than liquid hydrocarbon fuels.

Stringent legislation in the USA (particularly California) is beginning to indicate a trend towards tighter control of fuel specifications, which also affects European markets. This may influence the future growth of the offshore oil and gas industry. Even assuming that demand for gasoline or diesel remains the same, tighter environmental specifications may lead to more complex and more expensive refining processes, which in turn may affect the relative values of different crudes. Some commentators see a trend towards vertical integration developing in the USA. The need to control the specification of crudes and to refine them more and more efficiently may limit the industry's ability to buy and sell crude cargoes on the "spot" market, and encourage once again the development of longer-term alliances between producer and refiner.[15]

It is not beyond the bounds of possibility that newer transport fuels and engines could begin to replace hydrocarbon fuels in some parts of the transport industry. Hydrogen fuel cells produce no polluting exhausts – their by-products are electricity and water vapour. Prototype vehicles running on these cells are scheduled to be tested by some major fleet users, such as British Airways, in 1996. At the moment, however, the dominance of oil is almost complete. New fuels face several difficulties in establishing themselves – for instance, the need to demonstrate safety and reliability. For private road vehicle use, they also need to be available at a wide range of outlets; for this reason it is most likely that newer fuels would first be introduced in fleets which operate over limited distances from refuelling points, such as city bus fleets or taxi fleets.

The Impact of Disasters

One issue which the Scenario Planning Workshop considered was the potential impact of major oil or gas related disasters on public opinion. Large-scale leaks and fires from pipelines in the former Soviet Union have been reported. The *Exxon Valdez* oil spill in Alaska is another well-known example of an environmental disaster. Large oil spills have taken place in European waters, and it is not inconceivable that they will occur again – although as Chapter 2 made clear, the offshore industry is not usually at fault.

The offshore industry tends to be "out of sight and out of mind", and the amounts of oil which it spills are small. Two trends, however, may increase the environmental risks. With the possible exception of one blow-out in the Norwegian sector, there have been no pollution incidents on offshore platforms which have seriously affected the general public. However, as explorers search for new areas, they are turning their attention to oil and gas fields close to the shores of the UK, such as those in Morecambe Bay or Lyme Bay. Even relatively small spills in such areas might result in pollution of the shoreline, or damage to local industries such as fishing or tourism. Secondly, the increased use of pipelines to bring oil ashore from the major producing fields, and the age of some of the lines involved, raises the spectre of a catastrophic leak close to the shore from a major line.

The oil and gas industries go to considerable lengths to prevent accidents and disasters of this sort, and as a back-up they train staff to handle any emergencies which may occur. Major incidents, with extensive safety or environmental impacts, tend to lead to tighter legislation and regulation which affect all companies in the industry, not just the one which had the accident. One of the factors which the workshop considered was the likely effect on the industry if there were a serious pollution incident in UK waters. The reaction to such an incident could significantly alter the rules under which the industry operates – both the formal rules imposed by regulators and the informal rules imposed by society's expectations. For instance, if the oil- and gas-producing industries are widely seen as irresponsible polluters, they can hardly hope to recruit the best new talent from universities and schools. Nor can they expect an impartial hearing by governments, or local environmental tribunals.

Any discussion of environmental impact in the mid-1990s has to include the abandonment question, after the "Brent Spar" affair which was described in Chapter 4.

The Abandonment of Offshore Installations

The Thirteenth Driver: the Abandonment of Offshore Structures
How can this be done, what will the costs be, and what will the environmental implications be? In a favourable future, we believed abandonment decisions would not be rushed, but alternative uses of the infrastructure would be considered. A consensus would develop between industry and the public/government on how any abandonment that does take place is to be handled in an environmentally

— continued —

> *continued*
>
> optimal fashion. At worst, there would be premature abandonment, before the benefits of the existing infrastructure (e.g. for the development of nearby fields) had been fully appreciated. The abandonment process would be politically controversial and expensive and would consume disproportionate amounts of management time and company resources; it would distort public and government perceptions of the industry.

At the moment, any discussion of the environmental aspects of the North Sea is dominated by abandonment and the saga of the "Brent Spar". This affair raised some very interesting questions. It was a considerable victory for Greenpeace: having set itself the target of changing Shell's mind, Greenpeace used dramatic tactics and raised sufficient support to do so. The pressure was not placed on the British government; in effect Greenpeace and its allies went over the heads of the government by appealing to other European peoples and governments, and putting Shell under such pressure that it was forced to present the British authorities with a unilateral climbdown.

The result has forced all operators to rethink their abandonment plans. Many had assumed, and the government had encouraged them to assume, that after any environmentally unfriendly materials had been stripped out the platforms, or at least their supporting structures, could simply be toppled where they stood. What is now in question is whether the British government can give such authorisation – and if not, who can?

Government policy had largely followed the International Maritime Organisation (IMO) recommendations, defining minimum water clearance requirements above the toppled remains (55 metres above any remains in the northern North Sea). It remains to be seen how many companies are willing to try this option, which is still in theory available. Some current abandonment plans assume its use.

On the one hand, Greenpeace's methods may have alienated many people. Not all companies are as vulnerable to public pressure as Shell. Many exploration and production companies do not sell to the public, or operate in the most vociferous European countries, and would be relatively unmoved by popular pressures. It should also be possible to take steps to prevent the occupation of platforms by future protesters. In addition, it is open to question whose environment is being protected. If there is any environmental risk from disposing of such facilities as Brent Spar, can this be any better contained in a British or Norwegian shipbreaking yard than on the Atlantic sea bed? And if not, then are people and their immediate environment being put at risk to prevent damage to a remote area which has little perceptible influence on human life (e.g. through food chains or contamination of drinking water or air)? There is a respectable consensus of opinion that sea dumping is the preferable option.

On the other hand, the oil companies are now aware of the damage that can be done to their interests and reputation by a sophisticated and media based campaign, backed by a floating TV station and sophisticated communications facilities. When the issues are not clear cut, pressure groups can present them in misleading ways and arouse genuine public feeling.

Much of the North Sea infrastructure may well remain. Platforms may find new uses and new occupants as they become homes for deviated wells exploiting small fields some distance away (see Chapter 6). Pipelines may recover oil from newer fields as the production from older ones falls away. But the fact remains that abandonment is a serious issue. There are many installations which are unlikely to be needed, and which cannot simply be left where they are, to decay gradually and unpredictably, becoming a danger to shipping and a risk to the environment. (Pipelines are less of a problem. They can be cleaned out, filled with water or cement, and left; this may in some cases pose hazards to the fishing industry, but is much less of a hazard to the environment and none at all to shipping.)

Before the Brent Spar incident, the UK Offshore Operators' Association put forward its official view.[16] It anticipated that between 1995 and 2005 about 50 installations would be abandoned, at a total cost to industry of £1.5 billion. (However, since much of the costs of abandonment can be offset against tax, it has been estimated that 60% of the costs will be borne by the British taxpayer.[17]) It identified a series of removal options. These include total removal of all parts of the structure, described as the "only method for shallow waters"; partial removal, only available in waters deeper than 75 metres; toppling, where jackets and topsides can be cleaned up and then felled like trees so that they lie on the sea bed; deepwater dumping; and onshore disposal, with the structure towed to a licensed disposal site, cut up, and disposed showing due "duty of care" for the onshore environment.

The economic arguments for these various abandonment methods are complex. Fixed platforms in UK waters may be divided into several groups. There are some 205 in total, of which 135 are in the southern sector, 36 in the central sector, and 23 in the northern Sector, with a further 9 in Morecambe Bay.[18] Over 150 out of the 205 UK platforms are in waters less than 75 metres deep, and weigh less than 4000 tonnes; they would therefore have to be removed entirely, and this has been clear since 1989 when the IMO guidelines were drawn up. The number of jackets weighing over 10 000 tonnes in UK waters is only 32 out of 196; the number of topsides over this weight is 38. Only 7 are in more than 150 metres of water. The UK also has only 9 concrete platforms, as opposed to 15 in Norwegian waters (16 once the Troll "A" has been installed).

In the Gulf of Mexico, from water depths down to 150 metres, 914 platforms were removed between 1987 and 1994. All have been removed to sea-bed level; some have been replaced elsewhere as artificial reefs. Most of these were small – roughly equivalent in size and water depth to the southern North Sea platforms which form two-thirds of UK platforms. Eight were in depths over 100 metres, which is much closer to the problems posed by the larger northern North Sea platforms.

In order to conform with the London Convention, any platform or topside which is to be sunk must first be stripped of any harmful substances. The defined "harmful substances" include many which are commonly used: e.g. persistent plastics such as ropes and netting, and crude oil waste. Permits are required to dump lead, copper and a variety of other materials which may be used in the construction or operation of platforms. To remove these items offshore and return them to shore is expensive. Once this has been done, however, it may well be the cheapest option to topple the platform and, if the water depth is adequate, leave it on the sea bed. However, the value of the scrap metal is lost. Removing the topsides module by module and returning them to the shore to be stripped and either prepared for reuse, or converted to scrap, is perfectly feasible. Cutting up the jackets and raising them piece by piece, again to be returned to the shore, is technically possible, but has yet to be proven. Advocates of this method of abandonment point out that all the jackets and topsides were taken out from the land, and therefore it should be possible to return them. However, there is a great deal of money to be made from the various solutions to the abandonment problem, and the arguments are heated.

One case history shows that the Greenpeace victory over Shell may have done some oil companies a good turn. One of the platforms scheduled for early abandonment was Hutton, in the northern North Sea. This appeared to be a classic candidate for decommissioning and disposal: an old platform on a declining field, becoming more and more expensive to operate and less and less profitable. It came on stream in 1980, operated by Conoco in a partnership that included Oryx. In 1995 less than a sixth of the potential oil remained in the field, and the platform was to be abandoned in 1996. Oryx, however, thought it might be possible to extend the economic life of the field by cost cutting and looking for possible satellite wells. Accordingly, Conoco sold its share in Hutton (and several similar ageing fields) to Oryx in 1994. In February 1995, Oryx's tactics paid off: using 3D seismic to re-examine the structures around the existing reservoirs, it discovered a new untapped area of oil which it now hopes to develop.

Abandonment is no longer an "easy option", and several operators are re-examining the lives of fields and the platforms in them. This may lead to a more efficient exploitation of the resources under the North Sea. The Conoco/Oryx deal is one pointer: it may be that some companies will specialise in squeezing value out of older fields, whilst others concentrate on finding and developing newer ones. In theory any widespread transfer of liabilities is open to abuse – owners of offshore assets in the USA, for instance, are not allowed to transfer liability for them unless the acquiring group shows suitable net worth and surety against future environmental problems.[19] The UK licensing system also allows the government to ensure that operatorships are only sold to companies which can meet the ultimate abandonment liabilities involved.

However, if the option of toppling platforms and sinking them where they are is removed, the costs of abandonment will almost certainly increase. Although much of the cost is met by the government in the form of tax relief, the ultimate future of offshore installations may still have a serious impact on the "bottom line" of many companies.

INDUSTRY REPUTATION

The Fourteenth Driver: the Reputation of the Industry
The oil and gas industries are both more glamorous and more suspect than most other industrial activities. In a favourable future,

———— *continued* ————

continued

the industry would be seen as socially responsible. It would attract bright young people to work in it. It would earn the trust of the public and regulators to manage its own affairs well and conscientiously – government and others would "listen" to industry and give its requirements a fair hearing. In the less favourable situation, it could suffer from a lack of new talent, and attract tighter regulations and public suspicion or hostility. This driver depends very much on the industry's own efforts.

The reputation of the oil and gas industry in the eyes of the public over the last century has been, to say the least, variable. Soon after the industry was born, Rockefeller built up Standard Oil into a near monopoly within the USA (and almost, for a time, of the world oil trade); his ambition and aggressive methods were a sinister start for the industry. Some of the more flamboyant "oilmen", such as Armand Hammer, have done little to discourage an industry reputation for ruthless wheeling and dealing. Past attempts by the "majors" to tie up the world market among themselves (such as the 1928 Achnacarry Agreement) have come to light. Until the development of large fields in developed nations such as the UK in the 1970s, almost all oil (USA domestic production excepted) came from the less developed world but was sold to and used by developed countries. Its production and sale were dominated by developed country transnational corporations. This led inevitably to allegations of exploitation.

The sheer complexity of oil company operations does not help: few employees have a clear and detailed view of the whole range of their company's activities. It is many times more difficult for an outsider to penetrate the bureaucracy and private language of a large "major". Debates on such emotive issues as transfer pricing can only be conducted by accountants. As we have seen, it is almost impossible to say whether a tax rate applied to an oilfield development is "fair" or gives a suitable rate of return to a government; so much depends on the details and the method of assessing them.

On the other hand, the oil and gas industries have been welcomed in Britain as economic saviours. In many other countries they are a major source of revenue; government agencies vie to attract them and the investment they bring. Some countries send government teams to travel the world offering acreage, as others send arms salespeople or industrialists overseas, in trade missions led by government ministers.

The operations of the offshore companies are seen as glamorous and exciting – for instance, the oil well fire and blow-out capping specialist, Red Adair, was the subject of a film starring John Wayne and is still, in the mid-1990s, possibly the only "oilman" that the average British man or woman in the street would know by name (other than fictitious characters, such as J R Ewing of the television series *Dallas*).

Oil and gas companies employ public and government relations specialists to present their case. They have access to the technology and skills required to produce videos, glossy magazines and technical newsletters to present their activities in the best light. (Much of the detailed information about field developments in this book is based on such company sources.) They make considerable use of external public relations specialists: for instance, the United Kingdom Offshore Operators' Association (UKOOA) has hired a PR company to present the industry case in the abandonment debate.

Public relations skills cannot gloss over real issues. Piper Alpha was a tragic disaster caused by poor safety policies. The Brent Spar incident was a serious problem caused by insufficient understanding of the public reaction to a decision. The best policy for the industry, it seems, is to act responsibly, and to be aware of the expectations society has of it. (This does not mean that it should meekly accept these expectations without question: it has every right to argue its own case.)

The industry's ability to attract the right young entrants must be seen against the background of demographic changes in Europe. All mature industrial countries' populations are ageing, though Britain's less than most. In 1950, about 11% of Britain's population was over 65; in 2020, this is expected to be over 16%.[20] The number of people in the industrial world will have fallen from one-quarter of the world total to one-fifth. There will be pressure on industrial countries to use their shrinking labour forces more effectively; this is likely to mean later retirement ages, further encouragement for women to enter or re-enter the workforce, and more flexibility. It may also mean that job-hunting becomes a "seller's market" once again, particularly for those with special skills.

The oil and gas industry is re-examining its own size and shape. As discussed in Chapter 7 it is tending to become smaller and to focus on "core competencies" and higher productivity per employee. (It is by no means unique in this. Many other industries have "downsized", and there is a growing scepticism about the process.) In moving towards fewer workers, more flexibility, and more training or retraining, the offshore oil and gas industry is keeping up with the times. Since it depends heavily on scientific and engineering expertise, the offshore industry has always been heavily oriented towards training, and the larger companies at least have a good record. Being international gives it a further advantage: it can look for talent in countries with a larger labour pool. The vulnerability of the industry is that, with its high profile, its hazardous products, and its deep reach into almost all economic activity, it can easily alienate potential new employees by insensitive policies. Secondly, it is notoriously cyclical.

It might also be argued that changes in job structures within the industry will make careers less attractive. The graduate who at one time looked for a "safe", life-long career with one of the majors is unlikely to be offered it today. (But then, many people who thought 10 years ago that they had such safe careers now know that they didn't. Perhaps it is better to be realistic than comfortable.) Secondly, the oil industry is not greatly different from many other employers, who also find themselves unable to guarantee 40 years' future work.

We may see the breakdown of strong single-company loyalties. There are many employees of "major" oil companies, even today, who have never worked for any other company. (This is one of the main differences between the "majors" on the one hand, and the independents or most companies in the service sector on the other. It gives a unique flavour to dealings with some employees of the "majors".) If future employment patterns include more freedom to move – such as portable personal pensions rather than inflexible company schemes – and more incentive to move on as employers continually restructure, then individuals will have to take a broader view of their own loyalties. It is possible that many people with transportable skills (such as accountancy or personnel or some types of engineering skills) may move in and out of the industry, and bring wider perspectives.

Most of the technological and commercial developments outlined in preceding chapters have two things in common. They need fewer people, but those who remain to operate or manage must be far better trained and of high calibre. If oil companies cannot attract and continue to have access to such people, some of their triumphs may turn into disasters. This would be a shame.

This book is clearly written from an industry viewpoint. The author is a director of a small consultancy company which includes

216 *Waves of Fortune*

many oil and gas companies among its clients. In his experience, the industry can usually be proud of its record. It is no more secretive, no more exploitative, no worse in its safety and environmental policies and practices than any other industry. Because all its players are relatively large companies, constrained by legislation and active industry organisations and aware of the benefits of being good corporate citizens, it is often far better than other industries. It is also, as this book has tried to show, inherently fascinating. It provides an intellectual, technical and managerial challenge second to none. For this reason it would be a shame if it failed to attract the right sort of talent. It could easily become a backwater, staffed by corporate time-servers with little interest in further development or growth, stuck on a plateau at an altitude reached somewhere around the turn of the century.

This chapter and the preceding three examined the various "drivers" which may affect the future of the UK offshore industries. It is now time to move on to the four scenarios created by the original Institute of Petroleum Scenario Planning Workshop, which are set out in the next chapter.

REFERENCES

1. Van Meurs and Associates, quoted in *Oil and Gas Journal*, 24 April 1995, p 78; and Petroconsultants, quoted in *Oil and Gas Journal* 3 April 1995.
2. *Oil and Gas Journal*, 16 May 1994, p 73.
3. *Oil and Gas Journal*, 3 April 1995, p 47.
4. Ranger press release, April 1993.
5. *Offshore*, August 1995, p 31.
6. Quoted in *Euroil*, April 1994, p 36.
7. See survey of activity in western Europe by Connie Hughes in *World Oil*, August 1994, p 57.
8. See J D Davis, *High Cost Oil and Gas Resources*, Croom Helm, 1981, p 219.
9. See C Harvey, *Fool's Gold*, Hamish Hamilton, 1994, p 211. Tony Benn called the Commission's intervention "outrageous".
10. *EC Energy Monthly*, January 1994, pp 61–9.
11. Kevin Leydon of DGXVII, speaking in April 1995, quoted in *EC Energy Monthly*, 16 May 1995, pp 77–8.
12. These figures are taken from DX XVII 1993 Annual Review, *Energy in Europe*, published in 1994 by the European Commission.
13. See article in *Euroil*, December 1995, p 42.
14. Figures from DG XVII 1993 Annual Review, *Energy in Europe*, published by the European Commission.
15. P Fusaro, *The Impact of US Environmental Laws*, quoted in *Petroleum Review*, May 1995, p 207.

16. Quoted in *Petroleum Review*, April 1995, p 176.
17. This estimate is by Mr Eggar, quoted in *Petroleum Review*, March 1995, p 116.
18. Figures and analysis based on a paper given at a conference hosted by the Institute of Petroleum on 16 February 1995 by Mr Hugh Williams, the Business Development Manager of Heerema UK Ltd.
19. *Offshore*, September 1994, p 19.
20. Figures from H Macrae, *The World in 2020*, HarperCollins, 1994.

10
Four Scenarios for the Future of the North Sea

The following four scenarios were developed by the Scenario Planning Workshop at the Institute of Petroleum in 1995, and subsequently published in a short technical pamphlet by the Institute.

Each scenario is written as if the year were 2010. Each is written from a different viewpoint: an investment bank report, a speech by the Minister of State for Energy to the Institute of Petroleum annual Oil Week dinner, a Parliamentary Select Committee report on the oil and gas industry, and a report and assessment in the *Financial Times* of the annual general meeting of a large British oil company.

All companies and events mentioned are fictitious. In particular, the company UK Oil & Gas plc, which figures in each scenario, is not a thinly veiled disguise for any one company or any group of companies: it is a "typical" offshore oil and gas production company. Its varied fortunes in the four scenarios are meant to be representative of the whole.

SCENARIO ONE: "TRIUMPH OVER ADVERSITY" (HIGH INDUSTRY PERFORMANCE/UNFAVOURABLE ENVIRONMENT)

Despite the unfavourable conditions of the early twenty-first century, the UK oil and gas industry is emerging as a reliable if dull performer, reports investment bank Smith Jones Brown (SJB) in its research analysis bulletin for January 2010. The year 2010 will see oil prices remaining at $10 per barrel (in 1995 money) with gas prices still around 8 pence per therm. High interest rates, and the attractive oil and gas prospects elsewhere in the world, have meant that the industry has had to become considerably more efficient in order to attract the risk capital it requires to sustain its exploration and development activities. It has risen well to this challenge, restructuring itself considerably over the last 15 years to do so.

The majors have further diversified their investments overseas, and are maximising short-term returns from their UKCS operations. Although the technology exists to recover up to 70% of oil and 90% of gas from many fields, it is not economic to do this at present price levels. But 4D imaging allows operators to recover up to 60% (oil) and 75% (gas) with low-cost techniques, while managing reservoirs sensibly to ensure that if the more advanced technology does become cost-effective, it can still be used. Many of the fields which were once slated for near-term abandonment are once again seen as future assets, and the structures and pipelines associated with them are still being maintained. To keep costs low, worn out topsides are being replaced by cheaper and simpler standardised packages. Where maintaining structures, which in some cases date back to the 1970s or 1980s, is too expensive, the new generation of FPSOs can provide cost-effective short-term recovery options. Only one or two of the major pipelines, now very expensive to maintain, cause real concern. If they fail, there is unlikely to be the capital or the will to replace them. This could lead to "holes" in the UKCS infrastructure net, where known reserves are no longer produced.

Some independents are taking advantage of the new technology to develop stand-alone marginal fields as small as 1–5 mmboe. Reusable floating production units, of many shapes and sizes, and standardised sea-bed completions have driven costs down, so that very small marginal fields can be profitable when there is infrastructure in place to support them – as for instance the 1 mmboe Sporran field, which is tied back through Forties. Infrastructure owners, keen to maximise revenues, are welcoming marginal developments which keep up their pipeline throughputs – although there is fierce bargaining over tariffs. A further innovation since 2003 has been the secondary market in acreage. Some majors have struck innovative risk sharing deals with independents, allowing the smaller companies to exploit niche opportunities for fields and acreage which until now have been locked away. As a result, UKCS exploration has continued, and production levels remain at 1.25 million barrels per day.

The industry was, of course, badly shaken by the Barents Sea pollution disaster of 2001. UK company spokespersons have shown, most recently to a European Parliament committee, that environmental damage of this scale could not occur in UK waters, but there is still a heavy burden of suspicion. Paradoxically, perhaps, some industry figures say that the tight new environmental legislation is actually a benefit in the longer term. Environmentally friendly operation, they say, tends to be more effective operation in the long run. Cost is less of a problem: the main needs are to change attitudes, rather than for the massive capital expenditure of the 1980s. But there is no doubt that the new regulations struck gloom over the sector as a whole, and share prices are now at modest levels. The oil and gas industries can at least pride themselves that their effective lobbying has secured a level playing field for energy, with other energy sources enjoying no significant advantage under the European Energy Tax that was introduced in 2005. The failure of attempts to liberalise the European gas market has reduced the market for UKCS gas, which is, of course, a growing share of UKCS reserves. If larger markets for gas cannot be found, a long-term question mark must hang over this sector.

The most successful companies, concludes SJB, are those which recognise that the oil and gas industry has changed completely from the boom years of the twentieth century. The key to success is tight control of costs, and this is achieved by focused research, cooperation with suppliers rather than adversarial relationships, and a continual willingness to adapt company structures and sizes to suit the twenty-first century. New risk-management techniques and risk sharing have made it easier for smaller firms to exploit niches. The recent round of mergers and acquisitions has demonstrated, however, that institutional shareholders look for evidence of operational efficiency, clear strategic objectives, and management teams of the highest calibre. In these difficult times they will only support companies that can produce this evidence. UK Oil & Gas's decision in 2009 to sell off its entire UKCS business to a consortium led by BTR was a warning that there are no sacred cows.

One of the greatest advantages of the UK industry is the ending of the protectionist attitudes and policies that we saw at times during the late twentieth century. British firms have now learned to compete, and can offer the oil and gas industry an efficiency and cost-effectiveness second to none – both in opening up new areas, and in sustaining production under adverse cost environments.

This sector remains dull, but SJB believe there is reasonable medium-term value for investors in the better companies.

SCENARIO TWO: "THE BEST OF BOTH WORLDS" (HIGH INDUSTRY PERFORMANCE/FAVOURABLE ENVIRONMENT)

Extracts from speech by the Minister of State for Energy at the Institute of Petroleum Annual Dinner, 19 February 2010.

Madame Director General, ladies and gentlemen. I am happy to be able to address this eminent gathering at the end of 15 of the most successful years the United Kingdom oil and gas industry has ever known. After the volatility of the early 1990s, confidence grew that the oil price would remain steady at around $25 a barrel, and the gas price at around 20 pence per therm (both in 1995 prices), and so it has. This, and a range of new risk management tools, has allowed long-term investment decisions to be made with a high level of confidence. What began as a Wild West industry has matured and become more confident, and I think more socially acceptable. Your industry recognises that it needs this planet as much as anyone else, to inhabit and enjoy. I'm pleased to say that my own daughter, who is from a generation much more conscious of the environment than most of us here tonight when we were in our twenties, is proud to be a drilling superintendent with UK Oil & Gas. It could so easily have been a different story.

I am also pleased to say that successive British governments have recognised the importance of the UKCS for Britain, and have worked closely with industry and other interested bodies to ensure that regulation and taxation decisions are taken after good cost–benefit analysis. It used to be thought that safety and environmental care were one side of a zero sum game: if the oil industry paid attention to them, its costs would rise and its profitability fall. I think the government can claim that our non-prescriptive style of regulation has helped to expose this myth. We have asked you, the industries, to say how you would meet the standards set in Brussels and Westminster. You have risen to this challenge with a series of sensible and innovative solutions. The Chairman of UK Oil & Gas said to me earlier that he believes you have found that most of the environmental challenges that government has set you have led to more efficient operation, rather than higher costs.

I'm afraid that there will always be taxes; but we recognise the contribution that the UKCS makes to the UK economy, and we believe that tax policy should continue to take a long-term view. So much so that I am delighted to remind you that for 10 years now we have had an absolutely stable oil fiscal regime, with not so much as a minor change to the legislation since the final abolition of PRT in 1999.

Your research and development record, too, has been excellent, particularly for smaller companies. It is now possible to drill a 10 000-foot wildcat well for $2 million. New exploration techniques, such as 4D imaging, give a better than even chance that any exploration well will strike oil or gas. Thanks to your work on subsea multiphase pumping and metering, it is possible to recover a full well-stream from a sea-bed completion over 50 miles to the shore without the need for an onsite platform or floating production facilities. Where recovery directly to shore or to existing infrastructure is not possible, the new generation of FPSOs offers flexible and low-cost alternatives. All sorts of components are standardised and costs have fallen dramatically as a result. This standardisation has been made possible by greatly increased cooperation between companies. The industry employs far fewer people, but they are more highly skilled. The CRINE initiative in the last decade of the twentieth century cut cost levels by a third, and you have succeeded in cutting them by as much again in the first decade of this century.

The older fields, too, have benefited. Given today's prices, new recovery and reservoir management techniques allow you routinely to recover up to 75% of oil and 95% of gas economically. You have used the existing platforms – some of them 30 years old – as bases for the exploitation of marginal satellite fields, in many cases undertaken by independent companies. I am glad to say that the inflexibility of the 1990s is no more: acreage is traded regularly and smaller companies with different cost structures and innovative ideas are able to exploit smaller fields – as small as 1 mmboe – and to use the existing infrastructure to bring oil or gas back to shore. It is odd to reflect how, in the 1990s, we were able to contemplate the haphazard abandonment of this infrastructure, when in fact many of the structures and pipelines have become almost a part of the national wealth, often leased by companies who did not build them. Abandonment decisions are now taken on a broader basis. And, even where this infrastructure is not accessible, fields of 1 to 5 mmboe are now commercially exploited on a stand-alone basis.

As a result, UKCS oil production remains at 2 million barrels per day, capex levels at £3 billion in 1995 prices, and investment money is readily available for most projects. Proved reserves remain high, even though most are in a myriad of smaller fields. It is alas almost certain that there are no more major fields waiting to be found, like those discovered west of Shetland in the late 1990s. However, demand has remained high, as European liberalisation has created new markets for UK gas, and the security of supply which the UK offers has been demonstrated to our EU partners. The series of environmental disasters that paralysed FSU gas production in the late 1990s, and the delays and uncertainty from which North African gas producers are only now recovering, have made UKCS gas a major component of the high European growth of the last 15 years. The UK supply industries have contributed to this success, and have also carried methods and competitive costs from the UK to win business in developments abroad.

Four Scenarios for the Future of the North Sea 225

SCENARIO THREE: "SUNSET INDUSTRY" (LOW INDUSTRY PERFORMANCE/UNFAVOURABLE ENVIRONMENT)

The House of Commons Energy Select Committee Report into the UK Oil and Gas Industry, published in January 2010, assesses the present state of the industry. Compared with its peak in the late twentieth century, says the report, the industry is now in a sad state of decline, and the committee recommends further government support for the Aberdeen and Great Yarmouth regions which have been hardest hit by the collapse of UK Oil & Gas plc. The steady fall of oil and gas prices throughout the last 15 years – they are now at $10 per barrel of oil, and 8 pence per therm for gas in 1995 terms – has made it impossible for the UK's high-cost wells to compete. Only those fields which have already amortised their development costs, and which can be produced cheaply, are still in production. The UKCS now produces just under 750 000 barrels per day and this total is expected to fall much further by 2020.

The Committee examines the reasons for this decline in some detail. It accuses the previous British government, in 1998, of hastening the decline by imposing a new, ill-considered tax structure. Although this raised more revenue in the short term, its main effect was to hasten the closure of marginal fields and to shut down almost all the exploration activity needed to find future producing areas. One foreign operator actually walked away from its responsibilities, vacating a platform which was subsequently toppled by Royal Engineers at government expense. Other fields were abandoned in accordance with strict government regulations, and the Committee praises the DTI and Department of the Environment for their hard work in ensuring that environmental concerns were met to the fullest extent that technical and economic reality permitted.

The Committee fully accepts the government's responsibility for ensuring the safety and comfort of offshore workers, while recognising that some of the EU directives of the late 1990s may have made activities unnecessarily costly. Offshore work, it concludes, will always have an element of risk, and this should be accepted in any future legislation. It recommends, for instance, that the Minimum Working Hours Directive should no longer be applied to persons working offshore for less than 40 days in any year.

The Committee criticises the industry for its declining attitude towards safety and the environment in the past four years, and also for its lack of research and development. Several technical innovations developed in South-East Asia, India and Brazil have not been introduced: the world trend to simpler and lighter platforms has not been followed on the UKCS.

There were many corporate disasters in the sector. The Committee examined in particular the bankruptcy of UK Oil & Gas plc; this was partly ascribed to the reduced demand for hydrocarbons for transport purposes after the introduction of cheaper road taxes for the energy-saving "supercar", and inner-city road pricing concessions for electric vehicles across the EU and USA. Cheaper and more efficient coal-burning technology has cut down industrial gas usage.

However, the directors of UK Oil & Gas failed to adapt their business priorities and cost structures to meet the fall in demand, and after the disappointing results of exploration in the once popular area west of the Shetlands, and in "frontier" areas such as the Western Approaches, they made no further attempts to increase their reserve base in the UKCS. Among many poor business decisions, the directors should have foreseen that the Sporran Field FPSO contract, with McTavish Fabricators plc, was unworkable. The legal costs and damages awarded against UK Oil & Gas were to a large extent the final cause of the bankruptcy, despite the income stream from low-cost onshore acreage in Namibia which was UK Oil & Gas's main producing area. (Under a sale and leaseback deal, these fields belong to a Malaysian consortium; the sale proceeds were used to fund the last redundancy round in 2009, when UK Oil & Gas's staff were cut from 1250 to 800, and the number of offshore workers aged over 60 was halved.)

The committee is heavily critical of the Directors of UK Oil & Gas for retaining on the "back burner" much useful UKCS acreage which they had no intention either of exploring or of developing, but which might still have been economic as late as 1999 when there were several small independent operators listed on the UK Stock Exchange. The directors are also criticised for their "predatory" subsea pipeline tariffs, which drove two US independent operators out of the UKCS in 2008.

The Committee judges that there will always be a limited demand for an oil and gas industry on the UKCS, and recommends that the government should subsidise the remaining pipeline infrastructure, and pay the costs of "mothballing" some of the larger platforms, in case a strategic supply is ever necessary. New enhanced recovery techniques, developed in Kuwait, may give some of the major fields an "Indian summer" of production which could be called on stream at relatively short notice in times of international tension. This would be necessary if the UK is to have a secure source of jet fuel, for which there is still no alternative energy substitute. Because none of the remaining groups operating in the North Sea is British, the Committee recommends that field operators should, as a condition of future licences, be required to maintain extended storage facilities to help smooth any short-term oil price fluctuations, and that they should guarantee to transfer technology to British staff and suppliers. For the moment, however, the Committee believes that Britain's energy needs are best met by importing cheap oil and Russian gas, and by supporting the development of alternative energy sources.

SCENARIO FOUR: "GOLD PLATING" (LOW INDUSTRY PERFORMANCE/FAVOURABLE ENVIRONMENT)

Controversy about executive pay has never been greater, reports the *Financial Times* of 1 January 2010. Despite record low production and poor profitability, the Board of UK Oil & Gas (UKO&G) have increased their salaries by 175%. Justifying the increase, the Chairman pointed to the extensive R&D the company now carries out as "an investment in the future". Presenting the company's 2010 annual report yesterday, he claimed that UKO&G now has world leadership in oil dispersant technologies. He also highlighted a steady 1.7% earnings growth since 2007, and dismissed a private shareholder's claim that this is largely due to world currency movements and the steady price of oil and gas (equal to $25 per barrel/20 pence per therm in 1995 currency).

The Chairman reminded shareholders that the failure of two major subsea pipelines in the late 1990s, and the necessary environmental remediation and enhanced safeguarding, cut UKCS oil production by over 1 million barrels per day for nearly two years. Despite regular pigging and what he called "close monitoring" of "in retrospect, deceptively layered" corrosion, the previous management of the company had judged it right to operate these lines until they had burst, coincidentally within two weeks of each other, in March 1997, causing regrettable pollution to 100 miles of Scottish coast. Shareholders, he pointed out, would well recall the disastrous effects on the company's reputation and its share price. In an effort to restore both, the new managerial team took two measures:

- A rigorous programme of inspections has been introduced for every operation. UKO&G sets high targets, and it has consistently refused attempts to water these down to fit a "standard" system proposed by some other companies. UKO&G insists that its own staff audit every operation in which it has even the smallest interest. The company is proud to say that it has set a standard which others must now follow.
- all structures offshore, including platforms and pipelines, have been rigorously reviewed. By concentrating production and closing down some ageing fields earlier than planned, it has been possible to remove almost half of the company's offshore assets, thus minimising the risk of further disasters. The Chairman

reminded shareholders that both pipelines were successfully reopened in 1999 and have operated safely ever since. Capacity is back to a full 1.2 million barrels per day; although because of field decline, actual throughput is 0.6 million barrels per day and the line is expected to close in 2016.

Despite these successes, the Chairman conceded that the reputation of the oil industry was at a low ebb. Small independent companies had almost vanished from UK waters, as had many US companies. Two large companies, UKO&G and Euro-American, were often alleged to control the UKCS: they were accused of hoarding the remaining good acreage and removing the infrastructure which might have made the exploitation of marginal fields practical, or of denying its use to independents who with different approaches might have been able to work these areas profitably.

The Chairman deplored the low interest in the oil industry shown by graduates. It was not true, he said, that the industry had become complacent and unimaginative. New technology was under constant review in UKO&G's research centres, but would not be rushed into use before it had been properly evaluated. Azerbaijan's experience with floating production systems and full well-stream recovery over long distances direct to shore had shown that these technologies needed considerable refinement. He was proud to report on the success of the UKO&G staff suggestions scheme, which had led to significant cost savings in the new UKO&G Seabed Manganese Nodule Mining Division. (This division will build on UKO&G's considerable deepwater experience and diversify the company's asset base; he expected it to produce revenue by 2017.)

The oil companies could point to the success and security of their operations compared with those in the FSU, where major environmental disasters were an almost yearly event and production had virtually ceased, or to the Middle East, where a series of bitter civil wars between fundamentalists and modernisers had devastated the oil and gas industry in several key countries. About half of the UK's hydrocarbons were now imported from overseas (oil from West Africa and some products from East Asian refineries), but these countries were judged to be relatively secure sources. In return, UK gas was a major fuel in Europe; the UK as a result is a small net exporter of hydrocarbons.

He was proud, the Chairman said, that the UKCS produced a secure source of oil, safely and cleanly. Oil companies produced a steady stream of income, and the UK's supply and service industry had also grown to meet their needs: Britain now provides some of the best quality environmental technology in the world.

"Lex" Comment, from the *Financial Times*

UKO&G's statement does little to restore the reputation of a company which for over a decade has been trying to repair its damaged environmental record by a programme of research and diversification that has gone further and further away from its traditional core activities. It is little comfort to the investor that the government picks up a good part of the cost through "green" deductions from the tax take. Some old-fashioned attention to shareholder value would benefit stockholders and taxpayers alike.

CONCLUSION

It is no part of the purpose of this book to pick one of the four scenarios and declare that it is the most likely to happen. The scenarios are a tool for thinking about the future and not four predictions.

The future of the UK's offshore oil and gas industries will continue to be of vital importance to the country. These resources should not be taken for granted; that is why this book has tried to look back to the problems caused by lack of national resources of hydrocarbon energy in the 1970s and before. British people under 35 can scarcely remember a time when the country was not largely self-sufficient in its major energy sources. Their children, however, may have to live with the same problem.

At the same time, the future is in our own hands. This is why this book has attempted to show how and why some of the decisions which shaped the development of the industry were taken, in the hope that future decision makers will be able to approach the issues with a full understanding of the implications of their choices. Future decision makers will not only be those people directly employed by the oil and gas industries; they may be consumers, regulators, or any others with an interest that affects the industries and the political or economic ability to make their voices heard.

Appendix I
UK Governments from 1964 to 1995

Lead responsibility for the UK offshore industry has fallen under a succession of ministries. These were:

- to October 1969: the Ministry of Fuel and Power
- October 1969 to October 1970: the Ministry of Technology
- October 1970 to January 1974: the Department of Trade and Industry
- January 1974 to April 1992: the Department of Energy
- April 1992 onwards: the Department of Trade and Industry

The successive senior ministers (and the administrations of which they were a part) were as follows.

Conservative Government of Sir Alec Douglas-Home, October 1963/October 1964

- F Errol, Minister of Power to October 1964

Labour Government of Harold Wilson, October 1964 to June 1970
Ministers of Power:

- F Lee, October 1964 to April 1966
- R Marsh, April 1966 to April 1968

234   *Waves of Fortune*

- R Gunter, April 1968 to July 1968
- R Mason, July 1968 to October 1969

Minister of Technology:

- A Benn, July 1966 to June 1970

Conservative Government of Edward Heath, June 1970 to March 1974
Ministers of Technology:

- G Rippon, June 1970 to July 1970
- J Davies, July 1970 to October 1970

Ministry of Technology reorganised under Department of Trade and Industry in October 1970. Secretaries of State for Trade and Industry:

- J Davies, October 1970 to November 1972
- P Walker, November 1972 to March 1974

Secretary of State for Energy

- Lord Carrington, January 1974 to March 1974

Labour Governments of Harold Wilson, March 1974 to April 1976, and James Callaghan, April 1976 to May 1979
Secretaries of State for Energy:

- E Varley, March 1974 to June 1975
- A Benn, June 1975 to May 1979

Conservative Government of Margaret Thatcher, May 1979 to November 1990, and John Major, November 1990 onwards
Secretary of State for Energy

- D Howell, May 1979 to September 1981
- N Lawson, September 1981 to June 1983
- P Walker, June 1983 to June 1987
- C Parkinson, June 1987 to July 1989
- J Wakeham, July 1989 to April 1992

Department of Energy reorganised under Department of Trade and Industry in April 1992. Minister for Energy (under the Department of Trade and Industry):

- T Eggar, April 1992 onwards

Appendix II
Major UK Legislation Relating to the Offshore Industry

The first list is of the major pieces of legislation, except those relating to tax. These are in chronological order, to show the development of UK policy. (Not all of these Acts still apply, or apply in full.)
The second list is of some of the main Statutory Instruments (SIs). SIs are made under powers given to the relevant Minister by an Act of Parliament: they are usually administrative rather than contentious issues of policy. Again, this list excludes any tax regulations.
Neither list is exhaustive. Much offshore activity is also affected by normal onshore legislation, such as the Health and Safety at Work Act. The lists are intended:

- to show the extent of government involvement in the industry, and the amount of legislation and regulation involved;
- to show the changing nature of these Acts and Regulations and the changing focus of government over the last 30 years.

ACTS OF PARLIAMENT

- *Petroleum Production Act, 1934:* to vest in the Crown the property in petroleum and natural gas within Great Britain; to make provision

for searching and boring for and getting of petroleum and natural gas.
- *Continental Shelf Act, 1964:* to make provision for the exploration and exploitation of the UK Continental Shelf, and to give effect to provisions of the United Nations Convention on the high seas.
- *Mineral Workings (Offshore Installations) Act, 1971:* to provide for the safety, health and welfare of persons on installations concerned with the underwater exploration/exploitation of mineral resources in UK waters, and for the safety of such installations.
- *Offshore Petroleum Development (Scotland) Act, 1975:* to provide for the acquisition by government of land in Scotland for purposes related to exploration/exploitation of offshore petroleum, and to carry out works and facilitate operations for this purpose.
- *Petroleum and Submarine Pipelines Act, 1975:* to establish the British National Oil Corporation, and to make provision with respect to the functions of the Corporation; to make further provision about licences to search for and get petroleum, and about submarine pipelines and refineries.
- *Energy Act, 1976:* to make further provision with respect to the nation's resources of energy, including controls on the production and use of natural gas.
- *Oil and Gas (Enterprise) Act, 1982:* to make further provision with regard to the British National Oil Corporation (hiving off its offshore exploration and production interests to a private company) and the British Gas Corporation (to make provision for the supply of gas through pipes by persons other than the Corporation).
- *Oil and Pipelines Act, 1985:* to set up the Oil and Pipelines Agency to manage some residual activities of BNOC, and for the subsequent dissolution of that Corporation.
- *Gas Act, 1986:* dissolution of the British Gas Corporation and abolition of its privileges regarding the supply of gas. New provisions to govern the supply of gas, including the setting up of the office of Director General of Gas Supply.
- *Petroleum Act, 1987:* to make provision in respect of abandonment of offshore installations and submarine pipelines, and for safety zones around offshore installations.
- *Offshore Safety Act, 1992:* to implement many of the recommendations of the Cullen Report on the Piper Alpha disaster.

– *Offshore Safety (Protection against Victimisation) Act, 1992:* to implement an additional Cullen recommendation, that safety representatives and others should be protected against victimisation.

STATUTORY INSTRUMENTS

1972
– No 1542: called for logbooks to be kept and deaths offshore to be registered.

1973
– No 1852: gave power to inspectors to visit and inspect offshore installations.

1974
– No 289: required installations to be certified by appropriate certifying authorities.

1977
– No 486: regulations about the life-saving appliances to be held on offshore installations.

1978
– No 611: regulates the fire-fighting equipment to be held onboard offshore installations.

1980
– No 1759: prohibits drilling and other related operations offshore unless a qualified drilling supervisor is in charge.

1981
– No 399: Diving operations at work regulations, specifying safety and record keeping requirements, etc.

1982
– No 1573: safety requirements for submarine pipelines.

1986
- Nos 1644/5: two orders specifying the division of national interest in two cross-border fields.

1987
- No 930: extends sex discrimination and equal pay regulations to offshore installations.
- No 2198: clarifies the criminal jurisdiction of police over offshore installations.

1988
- No 1213: provisions to alter length of licences "in order to take further steps towards ensuring that acreage is not left unworked for long periods."

1989
- No 971: makes provision for "the workforce" offshore to elect safety committees, and allocates functions and powers to safety representatives.
- No 978: implements Cullen recommendation that offshore installations should have adequate emergency valves on pipelines and risers.

1992
- No 1314: transfers functions of Department of Energy (which ceased to exist) to Department of Trade and Industry.
- No 2885: sets out the requirements for Safety Cases for all offshore installations and rigs.
- No 3279: implements European Union Directive 90/533/EEC – regulations on contracts for supply and works on offshore and related installations.

1994
- No 3246: applies the Control of Substances Hazardous to Health (COSHH) Regulations to offshore installations.

1995
- No 738: Offshore Installation and Pipeline Works Management and Administration Regulations, requiring health and safety objectives to be set and defining responsibilities and powers of those involved.

- No 743: Prevention of Fire, Explosion, and Emergency Response (Offshore Installations) Regulations: sets out requirements for emergency response plans and goal-setting approach to prevention systems.

Appendix III
UK Oil Production, Expenditure and Export Statistics

Year	Production (million tonnes)	Sales (£bn*)	Opex (£bn*)	Capex (£bn*)	Net exports (million tonnes)
1975	2	n/a	n/a	n/a	−90
1976	12	0.6	0.1	1.5	−86
1977	38	2.2	0.2	1.6	−53
1978	54	2.8	0.3	1.7	−43
1979	78	5.7	0.4	1.8	−20
1980	80	8.9	0.6	2.2	−6
1981	89	12.3	0.9	2.5	15
1982	103	14.4	1.1	2.3	28
1983	115	17	1.3	1.8	40
1984	126	20.6	1.5	1.8	48
1985	128	19.9	1.9	1.9	47
1986	127	9.3	1.7	1.8	46
1987	123	10	1.7	1.3	42
1988	114	7.3	1.7	1.5	29
1989	92	7.5	1.9	1.7	22
1990	92	8.7	2.3	2.6	4
1991	91	8	2.6	3.4	−2
1992	94	7.7	2.6	3.8	−1
1993	100	8.6	2.9	3.2	3
1994	127	9.5	3.0	2.5	29

Source: DTI *Energy Report*, 1995.
* all money figures are in "money of the day", unadjusted for inflation.

Appendix IV
UK Licensing Rounds 1964–94

Number	Year	Blocks offered	Blocks awarded	Terms in years
1	1964	960	348	6 + 40
2	1965	1102	127	6 + 40
3	1970	157	106	6 + 40
4	1971/2	421	282	6 + 40
5	1976/7	51	44	4 + 3 + 30
6	1978/9	46	42	4 + 3 + 30
7	1980/1	80	90	6 + 30
8	1982/3	159	70	6 + 30
9	1984/5	180	93	6 + 30
10	1986/7	127	51	6 + 30
11	1988/9	212	115	6 + 12 + 18
12	1990/1	161	107	6 + 12 + 18
13	1990/1	117	66	9 + 15 + extension
14	1992/3	435 + 49	104 + 6	6 + 12 + 18
15	1994	81	29	6 + 12 + 18

Note: A sixteenth and seventeenth round are under way at the time of writing.
Source: DTI *Energy Report*, 1995.

Glossary

ACOP Approved Code of Practice. In order to prevent legislation from becoming too detailed and too "prescriptive", some offshore legislation (such as PFEER) is issued with an ACOP. The ACOP is a commentary on, and expansion of, the regulations. It does not have the force of law, but if an accident happens the responsible company would be required to demonstrate that it had followed the ACOP, or prove that it had found and used a better alternative (see Chapter 3).
Block British LICENCES for offshore exploration and production are issued, usually during Licensing ROUNDS, for specific blocks. The UKCS waters are divided into quadrants, each 1° of latitude by 1° of longitude, and each quadrant into 30 blocks. Each block is numbered by its quadrant and then the block number within it. Thus the Ninian Central platform, for instance, is in Block 3/3. The average size of a complete block is around 100 square miles. See Appendix IV for a list of licensing rounds.
Blowout Uncontrolled escape of oil or gas under pressure from a well, usually resulting in a "gusher", or spectacular and highly dangerous plume of inflammable hydrocarbons. The only major North Sea blowout was from Ekofisk Bravo in Norwegian waters in 1977. See WELL CONTROL.
Capex (capital expenditure) The one-off cost of developing a new asset (i.e. building a new platform, drilling a well) or improving it (i.e. drilling a new well from an existing platform). As opposed to OPEX.
Controlled document In order to ensure that all copies of critical documents (such as SAFETY CASES or emergency response plans) are

up to date, they are issued in numbered copies to individuals, who are responsible for keeping their own copies up to date as amendments are issued. Sometimes referred to as "living documents", except by those who have tried to read them.

Crude oil A mixture of many different HYDROCARBONS and other substances, varying greatly, as found underground in RESERVOIRS. The density of crude is measured by its "API gravity": a high API number (e.g. 40) marks a light crude from which useful products such as gasoline can easily be REFINED; a low API number (e.g. 13) indicates a heavy crude, a thick tarry sludge of much less value. "Sweet" crude is free of the highly poisonous and corrosive gas hydrogen sulphide, unlike "sour crude".

Cuttings The rock, etc., removed from a well in the process of drilling it.

Cyclical Word used to describe any industry that is subject to "boom" and "bust" periods, especially any industry that builds excess capacity in the boom periods and then wonders what to do with it in the bust periods. Also used to describe any industry that goes through this cycle at regular intervals without apparently drawing any conclusions about the future. *Si monumentum requiris, circumspice.*

Department of Trade and Industry (DTI) British government Ministry charged with (among other things) formulating policy on the extraction of the UK's oil and gas, and with licensing and overseeing oil company activity. There was formerly a separate Department of Energy, which was absorbed into the DTI. The senior DTI Minister is the Secretary of State; his junior Ministers are Ministers of State, one of whom usually has special responsibility for energy; the Minister of State in turn may have a junior Minister called a Parliamentary Undersecretary of State, not to be confused with the Permanent Undersecretary of State and Deputy Undersecretaries of State who are senior civil servants. (See Appendix I for a list of Ministers.)

Downhole What it sounds like, though a grammarian might prefer "down the hole" (or well) – e.g. "downhole motors" turn the drill bit using a turbine inside the well rather than a "top drive" from the drill rig above the surface. "Hole" is much used in other expressions, e.g. "makin' hole". See DRILLING TERMS.

Downstream The refining, logistics and marketing side of the oil industry, as opposed to UPSTREAM. People who measure oil in tonnes. See UNITS OF MEASUREMENT.

Drilling terms There are innumerable technical terms used by drillers: e.g. drill strings, packers, hangers, liners, casing, whipstocks, mud motors, kellies, mouseholes and ratholes, etc. This book is not the place to explain them; like legal or medical jargon, their purpose is to separate the expert from the general public. Drillspeak comes in two generations. There are still massively built Americans around, busy "kickin' ass, pushin' tool and makin' hole". But the modern North Sea drilling superintendent is an earnest Scot in his early thirties, who speaks alternate sentences of acronyms and business school English (e.g. "We put an MWD tool in the BHA of our HTHP well. This shortened the payback period and maximised our return on the asset without altering the risk profile"). See TLAS.
E&P Exploration and production. The UPSTREAM activities with which this book is concerned.
Farm in/farm out Method by which one oil company assigns part of its share in a licence to another. Such transfers require UK government approval.
FPSO Floating production, storage and offloading unit. Usually a converted ship or rig which acts as producer, stores the production, and offloads it into a tanker. A cheap way of exploiting a small well.
Health and Safety Executive (HSE) British government body charged with ensuring the safety of workers offshore.
Hydrocarbons Organic compounds containing hydrogen and carbon. These vary greatly in weight and in their state at normal temperatures: some are gases, some liquids, some solids.
Independent A smaller oil company, usually operating only UPSTREAM. However, this term covers a multitude of companies: some (e.g. Amerada Hess) are very large players in the North Sea, and some are so small as to be little more than specialised investment trusts (see discussion in Chapter 4). Even quite small oil companies are large companies by other standards.
Jack-up A jack-up rig is a kind of MODU, suitable only for use in relatively shallow waters, which floats to its target area, and then raises itself above the sea surface by lowering massive steel legs until they touch the ocean floor and take the weight of the rig.
Licensing The British government, as owner of all the UK's oil and gas reserves, grants licences to companies, or more often groups of companies, to exploit these reserves. Licences are for specific BLOCKS and have a range of conditions, covering duration, activities and costs. See Chapters 2 and 9 for fuller descriptions.

Major A large oil company, which is "integrated" – i.e. with operations both UPSTREAM and DOWNSTREAM. Some companies (Shell, BP, Exxon) are undoubtedly "majors"; other integrated companies – e.g. Unocal – are large but not usually regarded as "majors". The "majors" are nowadays often equalled or exceeded in size by NATIONAL OIL COMPANIES, such as Kuwait Petroleum or Petroleos de Venezuela SA; but the national oil companies tend to conduct their E&P at home and none has more than a peripheral involvement in the UKCS.
MODU Mobile offshore drilling unit or "RIG"; see JACK-UP or SEMI-SUBMERSIBLE. A mobile platform which can be positioned above a sub-sea oil or gas field, from which to drill into it or control production. One of the great mysteries of the offshore world is the names that are given to MODUs. (Sonat Rather? Rowan Gorilla? Cecil Provine? What happened to the early names, like Ocean Explorer?)
MOPU Mobile offshore production unit. A MODU that has been altered to perform basic production functions, but not enough to be an FPSO. (Often used hopefully to describe an old MODU that can no longer earn its keep drilling.)
Mud Liquid used to lubricate the bit during drilling and to maintain WELL CONTROL. Oil-based mud (OBM) requires special handling to prevent pollution; newer muds use water or synthetic oils as a base. Mud also contains specialised chemicals which perform a variety of processes. (See Chapter 6.)
National Oil Company Usually used to mean the nationally owned company set up by an oil-producing country to own (and sometimes operate) its own national resources. Many, including Britain's BNOC, were set up in the 1970s; BNOC did not survive the 1980s but others, such as Kuwait Petroleum and Petroleos de Venezuela SA are among the world's largest oil companies.
Petroleum Greek for "rock oil". Anything relating to underground oil or gas. In Britain, "petrol(eum)" is used to mean fuel for cars; elsewhere this is confusingly called "gasoline".
Offshore installation manager (OIM) The person legally in charge of an offshore RIG or platform. Has defined powers similar to a ship's captain, including complete authority over a 500-metre radius exclusion zone around the installation.
Operator When a LICENCE is issued to a company or group of companies, one of them is legally responsible for the physical work of exploring and producing the block. This one is designated the operator, and undertakes all the work on behalf of the others.

Glossary 247

Opex (operating expenditure) The recurrent costs of operating a given asset: i.e. staffing, maintaining and running a platform, keeping wells producing, exporting oil via a pipeline. As opposed to CAPEX.
Petroleum Revenue Tax (PRT) A tax first imposed in 1979 and abolished in 1993 for new fields. The industry had come to like PRT so much that some companies complained when it was abolished, a phenomenon which only an accountant could explain fully (see Chapter 9).
PFEER The Prevention of Fire and Explosion, and Emergency Response (Offshore Installations) Regulations of 1995. (Despite their title, the regulations are not intended to prevent emergency response.) These regulations and their accompanying ACOP govern many aspects of safety offshore (see Chapter 3).
Produced water Oil or gas are often found mixed with water in a RESERVOIR, either naturally or as a result of water injection to maintain reservoir pressure. This water is recovered with the oil or gas through the wells, and is known as produced water.
Refine (refinery, refining) CRUDE OIL needs to be broken down into usable substances such as gasoline, diesel, kerosene, etc., before it can be used. The exact processes are outside the scope of the UPSTREAM industry, and hence of this book.
Reserves It is important to arrive at an accurate estimate of the quantity of HYDROCARBONS in a RESERVOIR. This is essential for UNITISATION, or to value the reserves (and thus the shares) of a company, or the assets of a country. It is also almost impossible. Different levels of confidence with which the UK government estimates its reserves are explained in Chapter 2. So far UK figures have usually proved to be underconfident: i.e. there was more than they thought. The same cannot always be said of the reserve estimates published in some company reports.
Reservoir Subsurface stratum in which oil or gas is found. It is usually a porous material in which the oil or gas is held under pressure, somewhat like water in a sponge. Like the water in a sponge, it is almost impossible to get it all out (see Chapter 6).
Rig Popular term, usually used for a mobile offshore drilling unit (see MODU). Calling larger fixed offshore structures ("platforms" or "installations") "rigs" is a sure sign of the novice.
Safety case Every platform or MODU operating in UK waters has to have a safety case, identifying (and quantifying) the risks involved in its operations and setting out means for overcoming these risks.

248 *Glossary*

The Safety Case must be agreed by the HSE before the operation can begin; once so agreed, it becomes the legal basis by which the operation's safety is guided (see Chapter 2). It is usually a CONTROLLED DOCUMENT.

Semi-sub(mersible)   A semi-submersible rig is a kind of MODU. It is built above two enormous pontoons or hulls. During transit it floats with these on the surface of the water. When it is operating over a subsea oil field, the pontoons are ballasted until the rig floats much lower in the water. The effect of the pontoons below the surface is to dampen out wave movements at the surface and make the semi-sub a much more stable platform for drilling, etc.

TLAs   Three letter acronyms (such as BHA for bottom hole assembly, BOP for blowout preventer, etc.).

UKCS   The United Kingdom Continental Shelf. Area of water around the UK over which the British government has complete rights to license and tax oil or gas exploration and production, though some Scots and some in Europe may dislike the word "complete".

UKOOA   United Kingdom Offshore Operators' Association: the industry association representing and funded by the larger oil companies.

Unitisation   Fields may extend over two or more BLOCKS which may be LICENSED to different oil companies. In this case, the size of the field is estimated, then the amount that lies within each party's block, and the rights to/revenues from the field divided up accordingly. This process is highly technical, inevitably rather speculative, and strongly affects the fortunes of all concerned.

Units of measurement   There are several different ways of measuring quantities of oil and gas. They can be measured as a flow or as a finite quantity, and oil can be measured by volume or by weight. Flow measurements for oil are usually in barrels per day (bd or bpd) or thousands (kbd) or millions (mbd) of barrels per day. Barrels are a measure of volume (one barrel equals 35 gallons). Finite quantities are normally measured by weight, in tonnes.

Converting one measure to another is not easy. Since the flow of an oil well can vary from day to day, it may be misleading to multiply the bpd by 365 and assume that this is the annual rate of production. Secondly, the weight of different crude oils varies; a barrel of one will be heavier than a barrel of another. For the sake of simplicity, however, a standard multiplier is used: it is assumed that 1 barrel per day equates to 49.8 tonnes per year, or that 100 tonnes per year equate to 2.008 barrels per day.

Gas measurements are by volume, but made more complicated because they are quoted in both metric and imperial units, as cubic metres or cubic feet (or standard cubic feet – scf). These units are used both for flow measurements (such as cfd – cubic feet per day) or absolute quantities (bcm – billion cubic metres).

Liquefied Natural Gas is measured in tonnes. Condensates are usually measured as if they were oil – i.e. in barrels, etc.

Gas and oil can also be measured in calorific terms: how much heat they produce. These units are not greatly used in the upstream world, but they are the basis for equating flows or quantities of oil and gas. Production is sometimes quoted in "barrels of oil equivalent" units (e.g. mboe: million barrels of oil equivalent), where both gas and oil are included and it is necessary to use a common unit. The usual conversion factor is that 1 million tonnes of oil equals 1.111 billion cubic metres (or 39.2 billion cubic feet) of natural gas. Or, one billion cubic metres of natural gas per year equals 890,000 tonnes of oil per year or 17,800 barrels of oil per day.

In this book figures have usually been quoted in the original units in which they were given.

Upstream The exploration and production part of the oil and gas industry, as opposed to DOWNSTREAM. People who measure oil in barrels. See UNITS OF MEASUREMENT.

Well control See Chapter 1 for a fuller explanation. Means of controlling the pressure within the well, to prevent a BLOWOUT and to enable oil or gas to be safely and reliably produced.

Wildcat A well drilled on a completely new prospect, and therefore with very limited ability to predict whether it will be successful (unlike wells drilled on a known prospect, to see how far a reservoir extends). "Wildcat" is used in company annual reports to gloss over the fact that the odds of success are a little better than roulette (about 8 to 1 against in the UKCS, as opposed to 35 to 1 in roulette), but the stakes very much higher.

WOW "Waiting on weather" before operations can start. A common occurrence in the North Sea and other UK waters.

Index

Abandonment 208–12
ACOP (Approved Code of
  Practice) 243
AEEU trade union 43, 156
Alliances, partnerships 149–53
Allied Chemicals 64
Allison, Roderick (HSE) 94
Alvarez, Al 68
Alwyn North field 76
Amerada Hess 76, 126
Amoco 76, 129
AMOSS 129
Anglia Field 88
Aran 79
Arco British 117
Argyll Field 56
Auk Field 63, 65

Bacton 2, 76
BALPA trade union 43
Benn, Tony 55–6, 58
Beryl Field 76
Boreholes Directive (of EU) 204
Blocks 243
Blowout 243
BP (British Petroleum) 2, 12, 18, 21, 26,
  53, 57, 66
  size and activities 44, 77
Brent Crude (as marker price) 169–70
Brent Field 60, 76
Brent Spar 91–3, 157, 208–12

Britannia Field 87
British Gas (formerly British Gas
  Council) 2, 23–4, 33, 159, 175–6,
  180
  size and activities 76–7
British National Oil Corporation
  (BNOC) 57, 65–6
Brown and Root 78
Burgoyne Committee 39
Burmah Oil 56

Capital availability 183
Capex 243
Carbon dioxide emissions 52
Carbon tax proposals 203
Chevron 61, 78, 85, 87
Coal use 20
Coastguard Agency 50
Common Data Access Ltd 124
Conoco 76, 86, 89, 212
Continental Shelf Act 1964 25
'Core competences' 159
Coiled Tubing (CT) 131–2
Corporation Tax 32
CRINE (Cost Reduction Initiative in the
  New Era) 144
Cullen, Lord 70, 71
Cullen report on Piper Alpha 70, 71
Cuttings 244
Cyclical nature of industry 244
Cyrus Field 147

Index

Dauntless Field 121
Davy/Bessemer field 128–9
Dawn Field 125
Denmark 35
Department of Trade and Industry (DTI) 36, 38, 82, 244
Deviated wells 17
Disasters 207–8
  (*see also* Piper, Ekofisk)
Discovery rate in UK waters 75
Drake, 'Colonel' Edwin 12
Drake, Sir Eric (BP) 24, 100
Drilling techniques 14
Dunbar Field 155

Eakring 20
Earnings in Grampian region 46
'Economic rent' 31, 33
Eggar, Tim 65, 75
Ekofisk Bravo 48–50
Emerald Field 77, 138, 160, 180
Emergency Response 19
Employment in offshore industries 38, 46
Energy Act, 1976 36
Energy Charter 202
Energy Consumption 205–7, 240
Enterprise Oil 76–9
Errol, Frederick 54
Esso 2, 21, 63, 76
  size and activities 77
European gas markets 177–8
European Union 199
Exxon *see* Esso
'Exxon Valdez' 72

Fay, Dr C (Shell) 81, 97
Female workers offshore 44
FLAGS 77
Foinaven Field 126
Former Soviet Union (FSU) 166, 178
Forties Field 24, 60, 76
4D seismic 122
FPSO 136–9, 148, 245
Frigg gas supplies 90, 177

Gaisford, Dr Rex (Amerada Hess) 142
Gallaher, Paul (AEEU) 156
Gas markets 175

Gas prices 23, 175
Gas production 165
Geological surveys 12
Getty Oil 64
GMB trade union 43
Goal Petroleum 79
Grampian Regional council 46
Green Party (UK) 47
Greenpeace 48, 92–3
Groningen gas field 21
Gryphon Field 138
Gulf Cooperation Council (GCC) 166
Gulf of Mexico 18

Hamilton Oil 12
Hammer, Dr Arnold 69, 213
Hardy Oil and Gas 79
Harvie, Christopher 46
Health and Safety Commission (HSC) 39
Health and Safety Executive (HSE) 40, 71, 245
Heath, Edward 53, 233
Heather Field 64, 84
Hedging oil and gas prices 170, 182
Hudson Field 125–6
Hutton Field 212
Hydrocarbons Licensing Directive (of EU) 200

Immingham Terminal 76
'Independents' 79–80
Industrial relations 41, 71, 84
Information brokers 153–4
Infrastructure access 189–90
Institute of Petroleum 4, 99
Interconnector gas pipeline 175–8
Internal Rate of Return (IRR) 31
International Maritime Organisation (IMO) 92
International Monetary Fund (IMF) 59
Iran 12, 21, 29

Jackup rig 18, 245
Jones, Aubrey 36
Judy/Joanne fields 146–7

Kemp, Professor A 143

Index   253

Kerr McGee   18
KPMG Peat Marwick   33
Kuwait   66, 168

Labour Party   25, 26
LASMO   76–9
'Learning companies'   158
Legislation in UK   235–9
Leman Field   21, 76
Liberalisation of UK gas market   176
Legislation (UK)   235–9
Licensing by UK   34, 197–9, 241, 245
LOGGS (Lincolnshire Offshore Gas Gathering System)   76, 89
Load factor   24

Mabon, Dr   60
MAFF (UK Ministry of Agriculture, Fisheries and Food)   50
Magnus Field   76
'Majors'   79–80
MAR regulations   94
Marker Crudes   169
Methane Emissions   52
Miller Field   76
Mobil   70
MODU   246
MPCU (UK Marine Pollution Control Unit)   50
Morecambe Bay   5
MSF trade union   43
MSR (Midland and Scottish Resources plc)   76–7
Mud   16, 51, 132–3, 246
Murphy Petroleum   78

NAM (Nederlandse Aardolie Maatschappij)   21
National Union of Seamen (NUS)   43
Nature Conservancy Council   50
Nelson field   76
Neste North Sea Ltd   78
Netherlands   23
Ninian Field   61, 77, 78, 83, 85–90
Nitrous oxide (NOx) emissions   52
'North Sea Oil'   5–6
'North Sea Stock'   35
Norway   74

'Ocean Prince'   18
Occidental Petroleum   64, 69, 180
Odell, Professor Peter   72, 173
Offshore Supplies Office (later Oil and Gas Projects and Supplies Office) (OSO)   47
'OGJ 300'   142, 180
Oil and Gas (Enterprise Act) 1982   66
Oil and Pipelines Act 1985   66
Oil futures trading   170
Oil Industries Advisory Committee (OIAC)   40
Oil Industries Liaison Committee (OILC)   43
Oil price   29, 55, 72, 172–5
Oil production statistics   164
Oil spills   49
OPEC (Organisation of Petroleum Exporting Countries)   29, 172
Operator   246
Opex   247
Orwell field   117
Oryx UK Energy Co Ltd   44, 78

Partnerships   63–4
'Petrojarl 1'   138
'Petrojarl Foinaven'   136
Petroleum and Submarine Pipelines Act 1975   56
Petroleum Production Act, 1934   20
PFEER   94, 247
Piper Field   60, 64, 68
Prevention of Oil Pollution Act 1971   51
Procurement Procedures Directive   201
Produced water   247
PRT (Petroleum Revenue Tax)   32, 60, 64, 66, 247
Public Accounts Committee   32, 60

Ranger Oil   73, 79, 84, 88, 184, 196
Reputation of oil industry   212–16
Reserves of oil   186–9, 247
Reservoir   247
Reynolds, George   12
Risley, Allan (Phillips Petroleum)   120–2
Risks   179

RMT union   43
Robens Committee   38
Rockall   25, 35
Rockefeller   12
Rome, Club of   47

Safety case   40, 71, 93–4, 247
SAGE (Scottish Area Gas Evacuation)   76
St Fergus   2, 76
Saudi Arabia   12, 18, 66, 168
Scenarios for UKCS   218–30
Scenario planning techniques   99–108
Schwartz, Peter   107
Scott field   76
Scottish National Party (SNP)   58
Scottish Office   50
Seabirds   49
'Sea Gem'   18, 21
Second World War   20
Seismic   12–14
Select Committee on Energy (of UK House of Commons)   37
Semi-submersible rigs   18–19, 248
Shell   2, 21, 26, 118, 183
  size and activities   63, 72, 77
  and Brent Spar   91–3
  use of scenario planning techniques   105
Smith, Charles (Chevron)   86
South Morecambe Field   76
Special Petroleum Duty (SPD)   32
Spot markets (oil)   170
Staffa Field   76
Standard Oil   12
Statoil   79
Stenching   22
Strathspey Field   84
Stretch, Jim (AGIP)   196
Sullom Voe   2, 76
Sunter, Clem (Anglo American)   105–7
Sun Oil Britain Ltd   78

Talisman   79
Tax   193–7
Tchuruk, Serge (Total)   163, 179, 183
Texaco   86
TGWU trade union   43
Thatcher, Denis   54
Thatcher, Rt Hon Margaret   42, 65, 69, 72, 234
Theddlethorpe   76
THERMIE programme   204
Thompson Newspapers   64
3D seismic   120–2
Total   76, 90
Town Gas   20, 22
Trade Unions   41
Troll field   118
Tuft, Vic (CRINE Secretariat)   146, 157

'Uisge Gorm' FPSO   131
UK Continental Shelf   6, 248
UK economy   44–7, 59, 73
UK government, future actions   192
UKOOA (United Kingdom Offshore Operators Association)   31, 43, 52, 60, 74, 96, 210, 214, 248
UK offshore reserves   186–7
Unitisation   248
Union Jack Oil   79
United Nations Law of the Sea Conference, 1958   25
Unocal   64
USA Oil/gas statistics   167

Varley, Eric   58, 233

Ward, Brian (Shell)   155
Wardle, Charles   74, 82
Wildcat wells   249
Wilson, Harold   25, 233
Wood Group   45
World hydrocarbon consumption   174
Wylie, Revd Andrew   68